# MANAGING OPEN SPACE IN SUPPORT OF NET ZERO:

Carbon sequestration opportunities and tradeoffs in the Alameda Watershed

FEBRUARY 2023

PREPARED FOR THE
**SAN FRANCISCO PUBLIC UTILITIES COMMISSION**

I0105734

PRIMARY AUTHORS
**Lydia Vaughn**
**Sean Baumgarten**

CONTRIBUTING AUTHORS
**Helen Casendino**
**Erik Ndayishimiye**
**Matthew Benjamin**
**Denise Walker**
**Clara Kieschnick**
**David Peterson**
**Gloria Desanker**
**Letitia Grenier**

PROJECT GUIDANCE
**Lydia Vaughn**
**Letitia Grenier**
**Erica Spotswood**

DESIGN AND PRODUCTION
**Ruth Askevold**
**Katie McKnight**
**Jennifer Symonds**
**Brandon Herman**
**Denise Walker**

SFEI
PUBLICATION #1118

## SFEI
San Francisco Estuary Institute

PREPARED BY **San Francisco Estuary Institute**

IN COOPERATION WITH AND FUNDED BY THE
**SAN FRANCISCO PUBLIC UTILITIES COMMISSION**

San Francisco
**Water Power Sewer**
Services of the San Francisco Public Utilities Commission

**SUGGESTED CITATION**

San Francisco Estuary Institute. 2023. Managing open space in support of net zero: carbon sequestration opportunities and tradeoffs in the Alameda Watershed. Funded by the San Francisco Public Utilities Commission. SFEI Publication #1118, San Francisco Estuary Institute, Richmond, CA.

**REPORT AVAILABILITY**

Report and technical appendix are available online at sfei.org/projects/alameda-watershed-carbon

# CONTENTS

# ACKNOWLEDGEMENTS

This project was funded by the San Francisco Public Utilities Commission (SFPUC). We are grateful to the SFPUC staff who contributed to the project through scientific support, data transfer, technical review, field tours, and photography, including Carla Schultheis, Ellen Natesan, Jessica Appel, Scott Simono, Mia Ingolia, Clayton Koopmann, Tim Ramirez, Jeremy Lukins, and others.

This project benefited from strong scientific guidance and thoughtful review offered by our Technical Advisory Committee (TAC). TAC members included Patrick Gonzalez (UC Berkeley), John Battles (UC Berkeley), Valerie Eviner (UC Davis), Felix Ratcliff (LD Ford, Consultants in Rangeland Conservation Science), Margaret Torn (Lawrence Berkeley National Laboratory and UC Berkeley), Alison Forrestel (National Park Service), Ronald Amundson (UC Berkeley), Lauren Hallett (University of Oregon), Elizabeth Porzig (Point Blue Conservation Science), and Maegen Simmonds (Lawrence Berkeley National Laboratory). We offer special thanks to Patrick Gonzalez for providing analyses of tree and shrub canopy carbon that were included in our carbon storage assessment. Additional researchers or practitioners who offered advice and support are Klaus Scott (California Air Resources Board), Allegra Mayer (UC Berkeley), Rebecca Ryals (UC Merced), Virginia Matzek (Santa Clara University), and Ian Howell (Alameda County Resource Conservation District).

Finally, we thank the many SFEI staff members who contributed to this project along the way, including Vanessa Lee, Gemma Shusterman, Robin Grossinger, Scott Dusterhoff, Alison Whipple, Kelly Iknayan, Ellen Plane, and Kendall Harris. §

# SUMMARY
## CARBON IN THE ALAMEDA WATERSHED

Photograph of woodlands in Alameda Watershed by SFEI.

# MANAGING CARBON IN OPEN SPACE TO OFFSET URBAN EMISSIONS

California's natural and working lands provide a broad set of functions for ecosystems and people, such as clean water and food provision, biodiversity conservation, climate regulation, and economic support for individuals and communities. Among these functions, **carbon storage and sequestration** in soil and vegetation has received increasing attention in the policy and management spheres, offering a **natural climate solution** that complements the deep cuts that are needed in fossil fuel emissions. To meet net zero greenhouse gas (GHG) emissions targets set by state legislation (AB-1279) and local climate action plans, decision makers and land managers are looking to California's open space—its forests, grasslands, shrublands, and wetlands—to help reduce GHG emissions and draw down atmospheric carbon dioxide ($CO_2$).

Given the urgency of climate change, the state of California and many local agencies have set ambitious timelines for GHG emissions reductions. For example, the San Francisco Climate Action Plan (City and County of San Francisco, 2021) calls for net zero emissions by 2040, defined as at least a 90% reduction in emissions relative to 1990 levels from transportation, building energy use, industrial processes, and other sources of fossil fuel emissions. To sequester residual emissions, the plan calls for increased carbon sequestration on natural and working lands. Offsetting 10% of San Francisco's 1990 emissions, or 800,000 million metric tons of $CO_2$ equivalents (MMT $CO_2$e) per year, would be comparable to taking 170,000 gas-powered passenger cars off the road (EPA, 2022). To meet these ambitious targets, municipalities in the Bay Area may look to large undeveloped public land holdings for carbon sequestration opportunities to offset residual urban emissions. To identify and evaluate potential carbon management opportunities, land managers and policymakers need accurate information about baseline levels of ecosystem carbon storage, potential carbon sequestration rates, and other advantages and disadvantages of proposed carbon management actions.

Numerous strategies—collectively known as **carbon farming**—have been proposed for land managers to enhance carbon sequestration on managed open space. In Mediterranean-type terrestrial ecosystems in the Bay Area, carbon farming strategies that land managers might consider include compost application on rangelands, riparian forest restoration, silvopasture (tree planting in areas with livestock grazing), cattle exclusion to promote woody vegetation growth, native grassland restoration, and open space conservation. These strategies vary in terms of the magnitude of potential GHG benefit, the timing and long-term durability of carbon sequestration, feasibility in different landscape settings, ecological impacts relative to other management goals, and other co-benefits and tradeoffs. Understanding the advantages and disadvantages of each strategy informs management decisions and can help planners set actionable, realistic targets for open space carbon management.

# HOW CAN THE ALAMEDA WATERSHED SUPPORT CLIMATE ACTION GOALS?

Encompassing 39,000 acres in the East Bay, the Alameda Watershed is one of the largest public land holdings in the San Francisco Bay Area. The site of several water supply reservoirs, the watershed is managed by the San Francisco Public Utilities Commission (SFPUC) for both water resources protection as well as biodiversity conservation, fire risk reduction, and other goals. The landscape is characteristic of the region, with rugged topography and vegetation dominated by grasslands, shrublands, and oak woodland habitats, along with riparian forests and wetlands (Summary Fig. 1). Carbon sequestration in the watershed's ecosystems has received increasing attention as a potential nature-based strategy toward San Francisco's net zero GHG emissions goal. To support SFPUC managers in making informed carbon management decisions, this carbon assessment for the Alameda Watershed offers scientific guidance on the watershed's current and potential performance as a natural climate solution. Two main objectives framed this analysis: to quantify existing carbon stocks in the watershed, and to evaluate opportunities to enhance carbon sequestration in the watershed's vegetation and soil.

Current levels of vegetation and soil carbon storage were estimated watershed-wide using a variety of data sources and models. Six potential carbon management strategies were then evaluated to assess potential carbon benefits and summarize co-benefits, tradeoffs, and other considerations. The analysis approach and management considerations presented in this study are expected to be applicable to many other open space settings around the Bay Area and throughout the central California coast.

Photograph of wildflowers, above, courtesy of Brian Sak, SFPUC

**Summary Table 1. Carbon storage within Alameda Watershed ecosystem types.**

| Ecosystem type | Area (acres and percent of total) | Vegetation Carbon Storage (MT C) | Soil carbon storage (MT C) | Total ecosystem carbon (MT C and percent of total) |
|---|---|---|---|---|
| Grassland | 12,744 (38%) | 15,400 | 586,000 | 602,000 (24%) |
| Coastal scrub | 1,815 (5%) | 11,100 | 116,000 | 127,000 (5%) |
| Chaparral | 4,777 (14%) | 64,500 | 306,000 | 371,000 (15%) |
| Oak savanna | 3,988 (12%) | 85,400 | 255,000 | 341,000 (14%) |
| Oak woodland | 9,557 (29%) | 320,500 | 612,000 | 933,000 (38%) |
| Riparian forest | 653 (2%) | 42,100 | 42,000 | 84,000 (3%) |
| **Total** | **33,534** | **539,100** | **1,918,000** | **2,457,000** |

Notes: For area and total ecosystem carbon, values in parentheses indicate the percentage of the watershed-wide total. Additional land cover types not evaluated in this assessment include water bodies and developed areas. For vegetation carbon, error values represent spatial variation in per-acre carbon stocks across 30 m pixels for a given ecosystem type. For soil carbon, error values represent the standard error from the data synthesis.

**Summary Figure 1. Distribution of generalized ecosystem types in the watershed.**

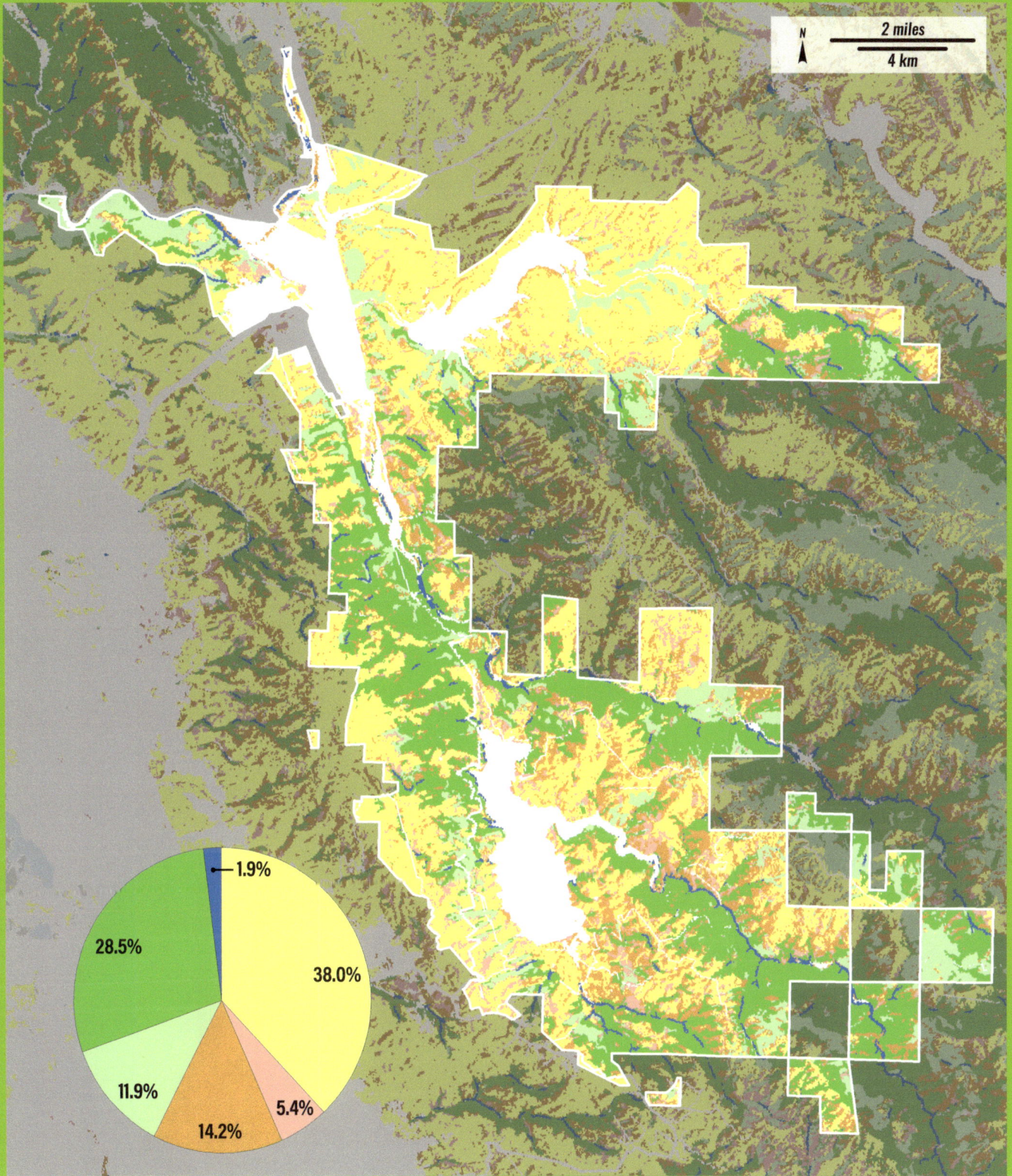

Notes: Ecosystem types were crosswalked from vegetation classes used in Gonzalez et al. (2015), based on 2010 LANDFIRE Expected Vegetation Type data. Because the composition of oak savanna is not well captured by the source imagery's 30 x 30 grid, vegetation community mapping by Jones and Stokes (2003) was used to define the extent of oak savanna.

**Ecosystem type**

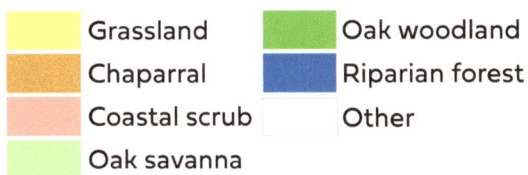

- Grassland
- Chaparral
- Coastal scrub
- Oak savanna
- Oak woodland
- Riparian forest
- Other

**RIPARIAN FOREST**

Vegetation carbon = 64.4 ± 11.8 MT C/acre
Soil carbon = 64.1 ± 6.3 MT C/acre

**OAK WOODLAND**

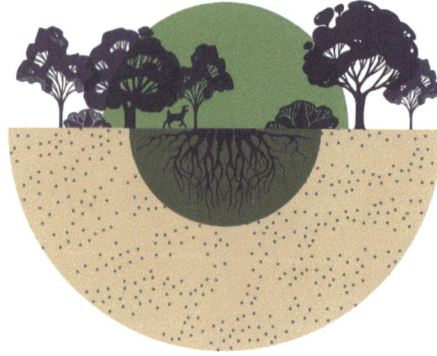

Vegetation carbon = 33.5 ± 10.8 MT C/acre
Soil carbon = 64.1 ± 6.3 MT C/acre

**OAK SAVANNA**

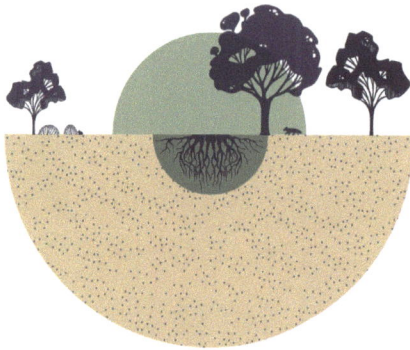

Vegetation carbon = 21.4 ± 10.8 MT C/acre
Soil carbon = 64.1 ± 6.3 MT C/acre

**CHAPARRAL**

Vegetation carbon = 13.5 ± 3.5 MT C/acre
Soil carbon = 64.1 ± 6.3 MT C/acre

**COASTAL SCRUB**

Vegetation carbon = 6.1 ± 0.25 MT C/acre
Soil carbon = 64.1 ± 6.3 MT C/acre

**GRASSLAND**

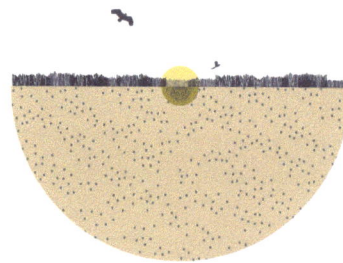

Vegetation carbon = 1.2 ± 0.14 MT C/acre
Soil carbon = 46.5 ± 3.9 MT C/acre

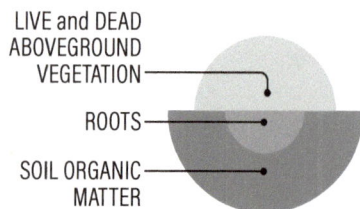

LIVE and DEAD
ABOVEGROUND
VEGETATION

ROOTS

SOIL ORGANIC
MATTER

**Summary Figure 2. Per-acre carbon storage by Alameda Watershed ecosystem types.** The size of each semi-circle represents the relative amount of carbon stored in vegetation and soil pools for each of the Alameda Watershed's six major ecosystem types.

# THE WATERSHED'S SOIL AND PLANTS STORE MILLIONS OF TONS OF CARBON

In total, ecosystems within the Alameda Watershed store an estimated 2.5 million metric tons of carbon (MMT C), including 0.54 MMT C in vegetation and 1.9 MMT C in soil (Summary Table 1). Vegetation carbon includes carbon stored in aboveground and belowground living plant tissues (including roots) as well as dead plant tissues such as standing and downed wood, leaf litter, and duff, while soil carbon includes carbon stored in soil organic matter.

Soil is by far the largest carbon pool (~4x greater than the vegetation carbon pool), though the proportion of carbon stored in soil varies from approximately 50% in riparian forest to 97% in grassland (Summary Fig. 2). Estimated soil carbon stocks are literature-based averages. Actual soil carbon storage varies according due to differences in soil type, vegetation structure, topography, and other variables. Vegetation carbon storage is lowest in grasslands and highest in ecosystems with dense woody vegetation such as riparian forest and oak woodlands; shrublands (coastal scrub and chaparral) and oak savanna have low to moderate levels of vegetation carbon.

While per-acre carbon storage is greatest in riparian forests, these forests occupy only 2% of the watershed area, and thus account for only 3% of total ecosystem carbon storage in the watershed. In contrast, grasslands account for 24% of total ecosystem carbon storage in spite of relatively low per-acre carbon storage, due to their extensive spatial coverage across the watershed.

Photograph by SFEI

**2010 Ecosystem
Carbon Storage**

>125 MT C/acre

0 MT C/acre

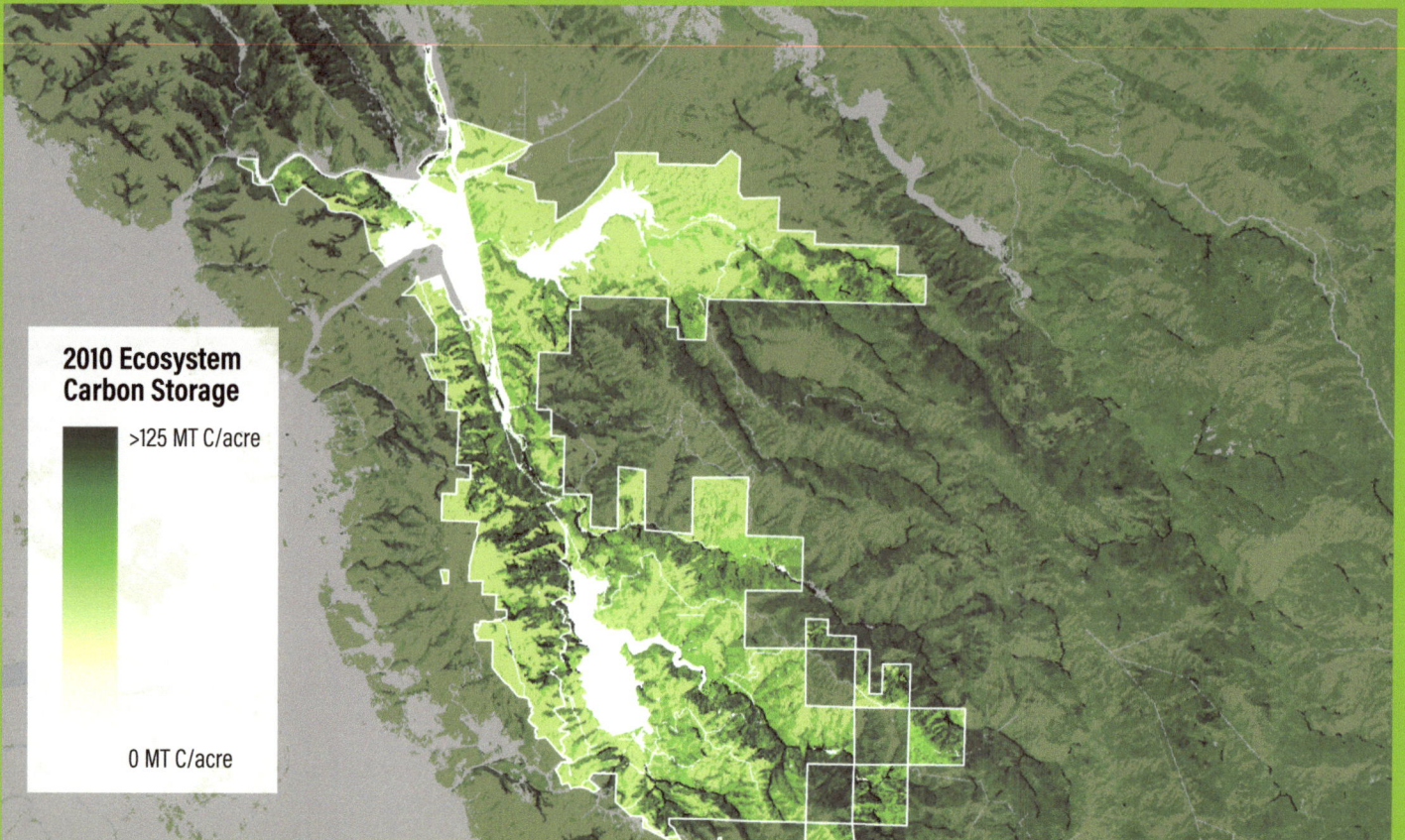

**Summary Figure 3. (top) Ecosystem carbon storage in the Alameda Watershed.** Carbon storage includes both vegetation and soils for sites not classified as water, barren, or developed. (bottom) Burn area in the Alameda Watershed of the 2020 SCU Lightning Complex Fires (in orange).

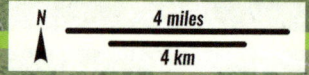

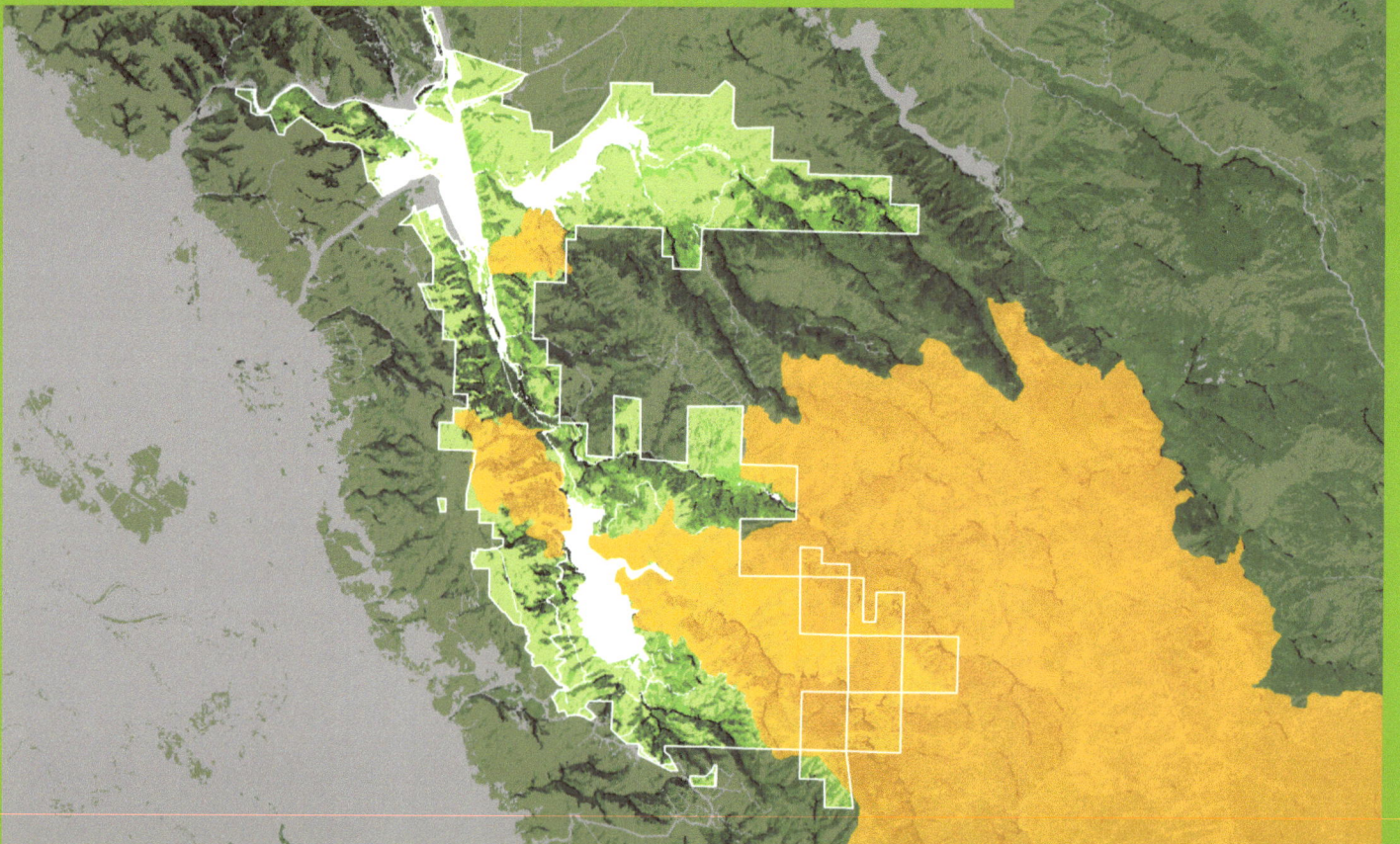

N

4 miles

4 km

# ECOSYSTEM CARBON MAY BE VULNERABLE TO WILDFIRE

Wildfire is common in Mediterranean-type ecosystems. Fires convert carbon stored in vegetation and surface soils into $CO_2$, methane, and other climate pollutants. In the Alameda Watershed, for instance, the 2020 SCU Lightning Complex fires burned 10,370 acres of watershed lands, resulting in an estimated loss of 33,100 MT C (~8% of total vegetation carbon; Summary Fig. 3). This carbon may be recovered in the coming decades through vegetation regrowth, but the pace and overall magnitude of recovery depends on regeneration success, vegetation succession, and potential future fires or other disturbances. California's 2020 wildfire season highlights the vulnerability of carbon sequestered in fire-prone ecosystems. If climate change increases wildfire frequency and severity, as has been predicted for California (Goss et al, 2020), carbon gains due to vegetation growth and regrowth may not be able to keep pace with wildfire-related losses (Gonzalez et al. 2015; Dass et al. 2018).

(below) Dropping fire retardant, Alameda watershed, (above) burned hillside, photographs courtesy of SFPUC.

Looking down at Alameda Creek, photograph by SFEI.

**Summary Table 2. Comparison of key considerations across carbon management strategies.** For a given consideration, green indicates strong support for the use of a management strategy, red indicates strong concerns, and orange indicates a low to moderate degree of support or concern.

| | Carbon and GHG benefits | Co-benefits | Tradeoffs | Feasibility |
|---|---|---|---|---|
| Rangeland compost | Low to moderate per-acre carbon benefits* | Likely benefits: forage production, soil quality, soil water retention | Key concerns: native biodiversity, residual dry matter control, water quality | Low to moderate concern: access to remote or steep sites |
| Riparian restoration | High per-acre carbon benefits** | Likely benefits: native biodiversity, soil quality, water quality | Moderate concerns: native biodiversity (including risk of pathogen introduction), water supply, wildfire risk, cattle water access | Moderate concern: tree establishment and survival, fencing maintenance, limited opportunity space |
| Silvopasture | Low to moderate per-acre carbon benefits** | Likely benefits: shading, soil quality | Moderate concerns: forage production, native biodiversity (including risk of pathogen introduction), water supply | Moderate concern: tree establishment and survival |
| Cattle exclusion | Low to moderate per-acre carbon benefits** | Potential benefit: water quality | Key concerns: agriculture, wildfire risk, native biodiversity | Moderate concern: fencing maintenance |
| Grassland restoration | Low, uncertain carbon benefits | Likely benefit: native biodiversity | Moderate concern: risk of pathogen introduction if container stock is used | Key concern: likelihood of restoration success |
| Open space conservation | High per-acre carbon benefits | Likely benefits: recreation, agriculture, native biodiversity, water quality, soil quality | Low concern: opportunities for alternate land uses | No major concerns |

**KEY** *Support for the use of a given strategy as a natural climate solution*

| | |
|---|---|
| 🟥 | Low benefits or substantial concerns |
| 🟨 | Moderate benefits or concerns |
| 🟩 | High benefits or low concerns |

*Assumes the material applied is composted green or animal waste. If biosolids are used, assumes material is amended to increase C:N ratios and limit $N_2O$ emissions

**Potential increased wildfire risk may decrease sustainability of carbon benefits.

# SOME "LOW REGRETS" MANAGEMENT ACTIONS CAN SEQUESTER CARBON AND BUILD ECOSYSTEM RESILIENCE

Land managers have a range of tools to increase carbon storage on natural and working lands, depending on the landscape setting and vegetation type. These carbon sequestration strategies vary considerably in their feasibility and potential to provide carbon and GHG benefits. For land managers focused on supporting healthy ecosystems, it is critical to assess whether carbon management strategies are aligned with other management goals such as biodiversity conservation and protection of water resources. A thorough consideration of co-benefits and tradeoffs is essential when managing for multiple ecosystem functions.

This study evaluated six potential carbon management strategies in the context of the Alameda Watershed, including **compost application on rangelands**, **riparian forest restoration**, **silvopasture**, **cattle exclusion**, **native grassland restoration**, and **open space conservation**. Summary Table 2 summarizes the key considerations associated with each strategy, including carbon and GHG benefits, co-benefits, tradeoffs, and feasibility.

Given the complexities, uncertainties, and tradeoffs inherent in decisions around carbon management, a prudent starting point is to identify "low-regrets" management strategies that sequester carbon with significant co-benefits and relatively few risks or potential negative impacts. The one strategy with very few risks is open space conservation, while strategies with minimal risks if applied in approriate settings include riparian restoration, silvopasture, and native grassland restoration. Cattle exclusion and compost application have significant potential tradeoffs that need to be carefully evaluated before these strategies are applied. Implementation of any carbon management strategy should begin with pilot studies and comprehensive monitoring to assess impacts on carbon and GHG exchanges, water quality and water budgets, ecosystem development, vegetation community composition, and other management priorities.

Photograph of Alameda Watershed by SFEI

# RESULTS AT A GLANCE

## WHY PRACTICE CARBON MANAGEMENT?

- **Protecting existing carbon stocks in forests, grasslands, and other natural and managed lands can help limit climate change and maintain healthy, resilient ecosystems.**

- **Management activities on these natural and working lands can tip the balance towards net carbon sequestration** by enhancing carbon uptake or reducing carbon losses.

- **Multi-benefit carbon management in the Alameda Watershed can set an example for other public and private lands.**

## CARBON STORAGE IN VEGETATION AND SOIL

- **Ecosystems within the Alameda Watershed store an estimated 2.5 million metric tons of carbon** (Equivalent to a year's emissions from 500,000 cars).[1]

- **This watershed's riparian forests and oak woodlands store as much carbon per acre as the Amazon rainforest.**[2] Grasslands and shrublands are less carbon dense, but account for a substantial percentage of the watershed's total carbon due to their large spatial extent.

- **80% of the watershed's carbon is stored belowground in soil organic matter.**

- The 2020 SCU Lightning Complex fires released around 8% of the total vegetation carbon in the Alameda Watershed. **As climate change continues, carbon gains due to vegetation growth may not be able to keep pace with wildfire-related losses.**

Photographs courtesy of SFPUC

1 EPA, 2022
2 Malhi et al., 2006; Moraes et al., 1995

# MANAGING CARBON IN OPEN SPACE

- **This study evaluated six potential carbon management strategies: compost application on rangelands, riparian forest restoration, silvopasture, cattle exclusion, native grassland restoration, and open space conservation.** Of these, open space conservation was the only strategy with high carbon benefits and few risks. All of the other strategies entail some tradeoffs and feasibility concerns, but may provide important co-benefits if implemented strategically.

- **If riparian restoration, silvopasture, and compost amendments were applied across all available space in the Alameda Watershed, carbon sequestration could offset as much as ~0.4% of San Francisco's 1990 greenhouse gas emissions.** This level of carbon management, however, could compromise other watershed functions and would likely encounter feasibility constraints.

- **Large areas of open space are needed to scale up the benefits of these carbon management practices.** To support climate targets, decision makers should look beyond this watershed's 39,000 acres for opportunities to conserve additional open space and practice multi-benefit carbon management on public and private land.

Photographs courtesy of SFPUC

# 1 INTRODUCTION
## CARBON IN THE ALAMEDA WATERSHED

Covering 39,000 acres around the Calaveras and San Antonio Reservoirs, the San Francisco Public Utilities Commission's (SFPUC) Alameda Watershed is one of the largest public land holdings in the San Francisco Bay Area. The Alameda Watershed landscape is characteristic of the Bay Area's East Bay hills, combining rugged topography, patchy woodland, open grassland, and scrub-chaparral. Managed by the SFPUC to maintain a reliable and safe water supply, the protected lands within the watershed support a range of other ecological and cultural values such as biodiversity conservation and rangeland forage production. Among these values, carbon sequestration in the watershed's vegetation and soils has received increasing attention as a nature-based strategy to mitigate climate change.

The value of natural ecosystems as sinks for atmospheric carbon dioxide ($CO_2$) is well recognized by scientists, governments, and land managers. Limiting global temperature increases requires not only reducing greenhouse gas emissions, but also removing $CO_2$ from the atmosphere, a process commonly referred to as "negative emissions" (Ciais et al., 2013; Gasser et al., 2015; Minx et al., 2018). Carbon sequestration in natural and working lands—in forests, in croplands, and in woodlands, grasslands, and shrublands such as those found in the Alameda Watershed—is one such means of negative emissions (Canadell and Schulze, 2014; Smith, 2016). Through the process of photosynthesis, plants absorb $CO_2$ from the atmosphere and store it in biomass and soils. This carbon resides in the ecosystem as living plant material, dead biomass, or soil organic matter until it is released back into the atmosphere through decomposition or fire. In this continual exchange, ecosystems can act as either sinks or sources for atmospheric carbon, depending on the balance between $CO_2$ uptake and emissions. Ecosystem conservation and management activities that enhance carbon uptake or reduce carbon losses are often referred to as natural climate solutions, as they leverage natural carbon flows to reduce atmospheric $CO_2$.

Carbon sequestration in natural and working lands is just one of many negative emissions strategies under development. Many such strategies are based in technology, including direct air capture (chemically extracting $CO_2$ from the air) and bioenergy with carbon capture and storage (Azar et al., 2010; Realmonte et al., 2019). In contrast, carbon sequestration in vegetation and soils is a nature-based solution that can, when implemented in an ecologically appropriate way, provide co-benefits for ecosystems and people (Canadell and Raupach, 2008; Huston and Marland, 2003). Ecosystem carbon management is often well aligned with other land management approaches aimed at ensuring long term sustainability of productivity, habitat quality, and other ecological functions (Follett and Reed, 2010; Horner et al., 2010; Lal, 2016). Riparian restoration, for example, can enhance bird biodiversity while sequestering carbon in vegetation and soils (Dybala et al., 2019); increasing carbon storage in rangeland and cropland soils can improve nutrient and water retention, increasing soil fertility and buffering productivity against drought (Ryals and Silver, 2013); and agroforestry practices can provide wildlife habitat, improve water quality, and increase carbon stocks in agricultural systems (Jose, 2009). In reality, both technology-based and nature-based negative emissions approaches will likely

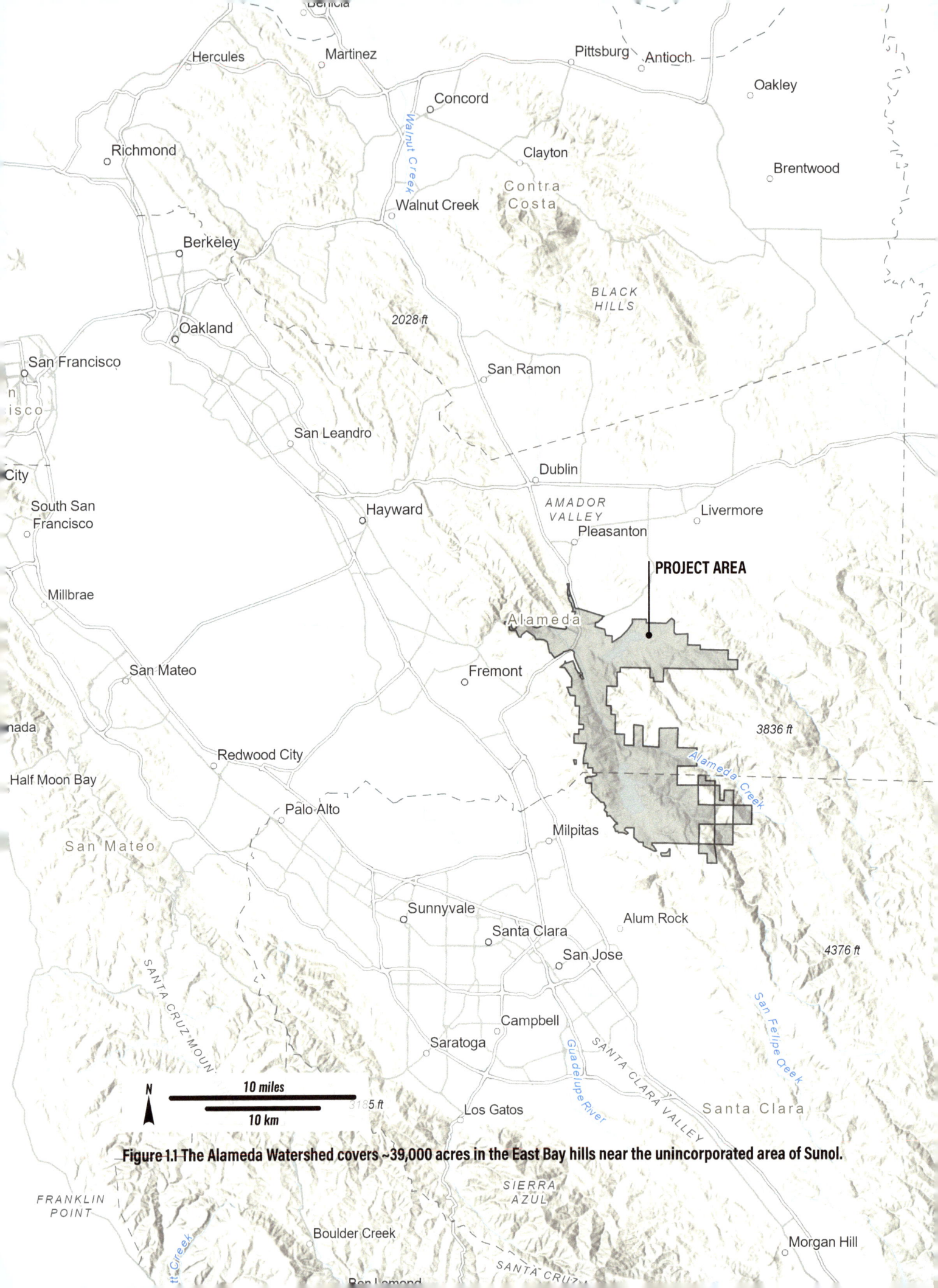

**Figure 1.1 The Alameda Watershed covers ~39,000 acres in the East Bay hills near the unincorporated area of Sunol.**

be needed to meet climate change mitigation targets. Within the portfolio of negative emissions strategies, protecting natural carbon sinks and increasing their capacity to sequester $CO_2$ can offer a range of benefits for ecosystems and people such as habitat conservation, enhanced soil health and fertility, improved water quality, and climate change resilience.

Habitat restoration in the Alameda Watershed, photograph courtesy of SFPUC.

In the oak woodlands, shrublands, and grasslands of the Alameda Watershed, potential synergies exist between carbon sequestration and ongoing management activities such as riparian restoration, low-impact grazing regimes, fire management, and native species conservation. On the other hand, land-based carbon management may present tradeoffs in certain circumstances with native biodiversity, water resource protection, fire risk management and other SFPUC management goals. Making informed decisions around ecosystem carbon management requires evaluating not only the potential magnitude of land-based negative emissions, but also the suite of synergies and tradeoffs. This type of analysis is timely and pressing. California recently codified the state's goal of achieving net zero emissions no later than 2045 (AB-1279), and the California Natural Resource Agency is required to set "an ambitious range of targets for natural carbon sequestration, and for nature-based climate solutions, that reduce greenhouse gas emissions...to support state goals to achieve carbon neutrality and foster climate adaptation and resilience" (AB-1757). The California Air Resources Board's (CARB) Climate Change Scoping Plan (CARB, 2022) calls out the importance of local governments in meeting these goals and urges local jurisdictions to adopt greenhouse gas emissions reduction targets and develop roadmaps to carbon neutrality. The San Francisco Climate Action Plan (City and County of San Francisco, 2021) is an example of such a roadmap, which calls for net zero citywide emissions by 2040. This plan defines net zero emissions as at least a 90% reduction from 1990 levels (8 MMT $CO_2e$/yr), and calls for nature-based solutions to sequester any residual emissions (up to 800,000 MT $CO_2e$/yr). To implement natural carbon sequestration at scale, the City and County of San Francisco may look to large public land holdings such as the Alameda Watershed for carbon management opportunities.

Clouds over the Alameda Watershed in spring, photograph courtesy of SFPUC.

To support SFPUC managers in making informed carbon management decisions, this report offers scientific guidance on the watershed's current and potential performance as a natural climate solution. This assessment was framed by two main objectives: to quantify current carbon stocks in the Alameda Watershed, and to evaluate opportunities to enhance carbon sequestration in its vegetation and soils. A central tenet of this analysis is that the value of any management action depends on the ecological context. Ecologically-thoughtful carbon management activities have the potential to support healthy and resilient ecosystems. In contrast, poorly considered carbon sequestration projects can degrade ecosystems or worsen climate-related challenges. For the Alameda Watershed, this ecological context is woven throughout the analysis, which addresses interactions between carbon sequestration, native biodiversity, water quality and availability, wildfire risk, and forage production on grazed lands.

## A guide to this report

This report provides basic information on the state of the watershed's carbon stocks and the science around carbon management so that SFPUC managers may make informed management decisions. This report consists of five chapters.

**Chapter 1 (this chapter) introduces the topic of carbon management on natural and working lands.** This chapter includes a very brief primer on carbon cycle principles and terminology used throughout this report.

**Chapter 2 offers background for the reader on how the ecology and management history of the Alameda Watershed influence its carbon storage.** Current carbon stocks, rates of $CO_2$ uptake and greenhouse gas emissions, and the potential for future sequestration all depend on characteristics of the ecosystem—its climate, vegetation, and soils—and the history of human activities on the landscape.

**Chapter 3 presents watershed-wide estimates of current carbon storage.** This chapter describes the modeling and data synthesis methods used to map carbon storage in vegetation and soils, and presents the findings with maps, graphs, and tables. This analysis provides context for evaluating future carbon management opportunities and quantifies the value of carbon storage that is currently provided by open space conservation in the watershed.

**Chapter 4 describes a set of potential strategies for managing carbon in the watershed.** For each management approach, this chapter provides an estimate of its carbon sequestration potential and presents key considerations for its suitability in the watershed. Such considerations include potential co-benefits for other ecological functions, potential tradeoffs with other management goals, and uncertainty and risk in its long-term capacity to sequester and store carbon.

Synthesizing findings from the previous chapters, **Chapter 5 offers broad recommendations related to carbon management.** This chapter does not prescribe specific actions, but rather highlights conclusions, discusses their implications, and identifies key questions and areas for future research.

# A BRIEF CARBON PRIMER

This report is written for a broad and scientifically curious audience. Analytical methods such as models, data synthesis, and uncertainty analyses are summarized at a high level throughout, with the goal of making the approach and findings accessible to a non-technical audience. For readers interested in greater methodological detail, an online technical appendix (https://www.sfei.org/projects/alameda-watershed-carbon) provides additional information on methods and underlying data sources. The following offers a quick primer on the basic science and terminology used in this report.

## Carbon stocks, carbon sequestration, and carbon sequestration potential

Grasslands, shrublands, forests, and other terrestrial landscapes store carbon in living vegetation, dead plant matter, and soil organic carbon. The amount of carbon stored in a given system is called a **carbon stock**. A carbon stock can be reported as the amount of carbon within a whole system (e.g., amount of carbon in the Amazon) or as a carbon density (e.g., quantity of carbon per acre). To represent carbon stocks, this report uses units of metric tons (**MT**), metric tons of carbon (**MT C**) or metric tons of carbon per acre (**MT C/acre**). The MT, equal to 1,000 kilograms, is a unit commonly used for carbon and $CO_2$ in the scientific literature and international agreements. For scale, a moderately sized valley oak contains around 0.5 MT C, and the General Sherman tree in Sequoia National Park contains an estimated 600 MT C.

These ecological systems are open systems. Carbon continually enters the ecosystem through photosynthesis and is released back into the atmosphere through decomposition and fire. When plants add carbon to an ecosystem faster than carbon is released, the ecosystem accumulates carbon. This process is **carbon sequestration**. In this report, carbon sequestration is defined as a rate of change in a carbon stock, in units of metric tons of carbon per acre per year (**MT C/acre-yr**) or metric tons of carbon per year (**MT C/yr**).

An ecosystem's carbon stock and carbon sequestration rate are not necessarily related in a straightforward way. The amount of carbon in a forest, for example, doesn't determine how fast that forest is gaining or losing carbon. It could be gaining carbon year to year through tree growth and soil carbon sequestration, or it could be losing carbon over time due to drought, pathogen outbreaks, fire, or other sources of tree mortality. Similarly, the present rate of carbon sequestration doesn't determine its **carbon sequestration potential**, or the amount of carbon that *could* theoretically be sequestered under a given management practice. Carbon sequestration potentials can be expressed as metric tons of carbon per year (**MT C/yr**), metric tons of carbon per acre per year (**MT C/acre-yr**), or a cumulative change in carbon storage over time (**MT C over a specified number of years**).

## Carbon, carbon dioxide and other greenhouse gases

The element carbon is integral to all living things, forming the molecular backbone of living organisms, dead and decaying organic matter, and fossil fuels. When these materials decompose or combust, carbon they contain is released to the atmosphere in the form of greenhouse gasses. The dominant form of carbon exchanged between ecosystems and the atmosphere is $CO_2$, but ecosystems also release other gasses and aerosols in smaller amounts. These can affect the climate as well, sometimes with greater potency than $CO_2$. One ton of methane, for example, has 28 times the greenhouse gas power as a ton of $CO_2$ when compared over a 100-year timeline (its 100-year Global Warming Potential, or GWP; Myhre et al., 2013). ($CO_2$ lasts much longer in the atmosphere than methane, so methane's GWP depends on the time horizon.) Reporting greenhouse gas emissions in units of $CO_2$ equivalents (**$CO_2$e**) accounts for these differences in global warming potential so that emissions of different greenhouse gasses can be added, subtracted, or compared. When managing ecosystems in order to promote carbon sequestration, it is important to consider these other potential influences on the climate. While this report focuses primarily on carbon stocks and potential carbon sequestration, it also identifies other non-$CO_2$ climate effects that should be taken into consideration.

For any management activity aimed at reducing the concentration of greenhouse gasses in the atmosphere, it is critical to think carefully about the bounds of the system. For this report, analyses are specific to the Alameda Watershed, with the potential footprint for management activities defined as the SFPUC land holdings around the Calaveras and San Antonio reservoirs. More broadly, however, management decisions in the Alameda Watershed may influence greenhouse gas emissions and other ecological effects over much larger geographies. Materials used for management, for example, may entail life-cycle greenhouse gas emissions prior to their end use, and habitat corridors may affect wildlife populations more broadly across their range. Where clear information is available, this report discusses life-cycle greenhouse gas emissions or large-scale ecological effects associated with potential carbon management approaches. §

Photograph by SFEI.

# 2

# ECOSYSTEMS AND ENVIRONMENTAL CONTEXT

## FOR CARBON MANAGEMENT IN THE ALAMEDA WATERSHED

In the Alameda Watershed, a combination of ecological and human factors determine the amount of carbon stored by the landscape and the potential to alter this carbon storage through management activities. Ecological factors such as soil texture and mineralogy, vegetation type and productivity, temperature, and the amount and timing of rainfall influence how carbon moves from the atmosphere into vegetation, from vegetation into soil, and then back to the atmosphere through decomposition (Bardgett et al., 2013; Chou et al., 2008; Davidson and Janssens, 2006; Sitch et al., 2003). Influencing these factors through thousands of years of land use, humans have modified the landscape's carbon dynamics through species introductions, fire, urban development, agriculture, and other anthropogenic disturbances. These processes and properties, both ecological and human, both biological and abiotic, define the context and landscape potential for ecosystem carbon management.

Biophysical factors influence carbon sequestration in landscapes both directly and indirectly. A site's vegetation and climate affect carbon sequestration directly by controlling biological rates of photosynthesis and decomposition. Indirectly, these rates are mediated by multiple interrelated environmental factors; soil texture, for example, interacts with vegetation, topography, and climate to determine erosion rates (Ravi et al., 2010), which affect the movement of soil carbon and its availability to decomposer organisms (Berhe et al., 2007; Gregorich et al., 1998). These biophysical factors affecting carbon cycling are often closely coupled with human activities, where ecological functions are influenced by human land use and management while guiding the way humans interact with the landscape. Three key factors controlling the carbon balance in the Alameda Watershed are the Mediterranean climate, mosaic of woody and grassland vegetation, and the history of human disturbance. These factors control present-day carbon storage, set bounds on the watershed's potential for carbon sequestration, and can guide us toward carbon management actions that have the greatest opportunity to significantly benefit the climate in the long term.

## MEDITERRANEAN CLIMATE AND ASSOCIATED FIRE REGIME

The Mediterranean climate of California's central coast is characterized by cool, wet winters and warm, dry summers, with high variability in the amount of rainfall among seasons and between years. Mediterranean-type ecosystems are found in five locations across the globe: California; the Mediterranean Basin; and regions of Chile, South Africa, and Australia. Relative to other ecosystems, Mediterranean-type ecosystems have moderate carbon density, storing an estimated 13 MT C/acre in vegetation (Gibbs and Ruesch, 2008) and between ~25 and 50 MT C/acre in soil organic matter in most locations (Batjes, 2016). In contrast, the high-productivity temperate oceanic forests of the Pacific Northwest store an average of 150 MT C/acre in vegetation (Gibbs and Ruesch, 2008); and Arctic tundra, where decomposition is limited by cold and frozen conditions, can store more than 160 MT C/acre in the top meter of soil (Gibbs and Ruesch, 2008). These large-scale patterns in carbon storage are determined by the balance of productivity and decomposition, which are controlled in large part by climate (temperature and water availability) and plant characteristics (Chapin III et al., 2009; Churkina and Running, 1998; De Deyn et al., 2008). In Mediterranean-type ecosystems like the Alameda Watershed, the climate, dominant vegetation, and disturbance regime favor moderate carbon storage in both vegetation and soils.

While climate and plant traits set the theoretical boundaries for carbon storage potential, local environmental disturbances such as fire and land management activities regulate actual vegetation and carbon storage patterns (Chapin III et al., 2009). Like other ecosystems with a Mediterranean climate, the shrublands, woodlands, and grasslands of California's central coast experience frequent fire. Much of the region's vegetation is adapted to withstand or benefit from fire (Davis and Borchert, 2006; Keeley et al., 2011; Safford et al., 2021; Sugihara et al., 2006), and the distribution of vegetation types across the landscape is influenced by wildfire frequency and severity (Keeley and Syphard, 2016). The effect of fire on carbon dynamics depends on the scale of observation. At the scale of an individual site, wildfire causes an immediate loss of carbon from the ecosystem to the atmosphere as carbon in living vegetation, dead plant material, and surface soil layers combusts and converts to $CO_2$. Over longer periods of time or larger spatial scales, however, fire is part of a continuous exchange between the atmosphere and the land surface. As ecosystems regenerate in the years after a fire, fire-related $CO_2$ emissions are removed from the atmosphere by new vegetation growth in the burned area. If these processes are in balance—if fire frequency, severity, and vegetation regrowth remain in equilibrium over time—wildfire emissions do not represent a net $CO_2$ source to the atmosphere (van der Werf et al., 2017).

In the ecosystems of central California, the incidence of fire has changed dramatically over the past two and a half centuries. Prior to Euro-American settlement, the Native peoples in the San Francisco Bay Area frequently used fire as a management tool. Periodic burning was used to enhance forage for wildlife, control pathogens, improve access to acorns, aid in hunting game, and for other purposes (Anderson, 2005). While the specific extent and frequency of historical burning within the watershed is unknown, a range of evidence suggests that grasslands and oak savannas in the region may have experienced a fire return interval of 5–10 years or less in many cases (Keeley, 2005; Rutherford et al., 2020; Safford et al., 2021). These frequent, low-intensity surface fires would have removed understory vegetation and dead plant material, reduced the density of oak savannas, and prevented conversion of grasslands to woody vegetation types. The pre-settlement fire regime of Diablan coastal sage scrub and other shrubland types in the region is not well understood, but would have been less frequent than in grassland and savanna vegetation communities and may have been similar to fire regimes documented in similar habitat types in coastal southern California (Cleland et al., 2016; Davis and Borchert, 2006; Keeley, 2006).

Cessation of Indigenous burning, Euro-American settlement and associated land use changes, and fire suppression practices have profoundly altered fire regimes in the region. Prior to the 2020 SCU Lightning Complex fires, which burned approximately 10,300 acres in the southern portion of the watershed, most of the watershed had not experienced fire during the period of modern record-keeping.[1] Many of California's oak woodlands now experience an estimated fire return interval of 55–70 years (Safford et al., 2021). Fire frequencies in Diablan coastal sage scrub and chaparral communities in the central coast region have likewise decreased in many areas (Davis and Borchert, 2006). Fire suppression in the region frequently results in the transition from more open vegetation types such as grasslands to more closed veg-

---

1     Fire perimeter records were obtained from the California Department of Forestry and Fire Protection (CAL FIRE). Records extend back to 1878, but are most complete for the period post-1950.

etation mosaics such as shrubland and woodland, though the presence of livestock grazing can mediate this effect (Callaway and Davis, 1993; McBride and Heady, 1968). Stanford et al. (2013) documented conversion of shrubland to oak woodland and then mixed evergreen woodland within the watershed, and attributed this transition primarily to the reduction in fire frequency.

The net effect of changes associated with climate change (changes in fire regimes, distribution of vegetation communities, etc.) on ecosystem carbon storage will likely depend on complex interactions between climate, fire, vegetation, and management practices (Batllori et al., 2015). Model projections for the region's changing climate include warming temperatures, particularly in summer, increasing frequency, magnitude and duration of heat waves, a small (uncertain) decrease in annual precipitation, and continued high interannual variability in precipitation (Cayan et al., 2008, 2012). Statewide, the incidence of autumn days with extreme fire weather increased more than twofold between 1979 and 2018, and such conditions are projected to increase substantially by 2100 in California (Goss et al., 2020). Furthermore, wildfire emissions in recent years have not been offset by vegetation regrowth: an analysis by Gonzalez et al. (2015) found that, between 2000 and 2010, wildfires accounted for approximately two-thirds of carbon losses from vegetation throughout California, and these emissions were not offset by carbon sequestration in new biomass (either in burned or unburned areas). With warmer temperatures, more frequent and longer droughts, and more extreme fire weather predicted for California, the risk of wildfire should be a guiding factor when assessing carbon management activities.

Wildfire in the Alameda Watershed, courtesy of SFPUC.

## MOSAIC OF VEGETATION

Two vegetation mapping sources were used to characterize modern vegetation mosaics in the Alameda Watershed. The LANDFIRE Program Existing Vegetation Type (EVT) data product, put out by the US Forest Service and US Department of the Interior, maps the dominant vegetation community at 30 m resolution according to Landsat imagery, vegetation structure data, and classification models. LANDFIRE layers for EVT, canopy height, and canopy cover were crosswalked to a set of more aggregated vegetation classes by Gonzalez et al. (2015) (Table 2.2), and these crosswalked vegetation classes were aggregated further for this analysis into six ecosystem types: grassland, coastal scrub, chaparral, oak savanna, oak woodland, and riparian forest (Table 2.1, Fig. 2.1). Because oak savanna is challenging to categorize at the 30 m grid scale, mapping conducted in the Alameda Watershed by Jones and Stokes (2003; Table 2.2) was used to define oak savanna areas not identified by the LANDFIRE-based classification.

Vegetation cover in the Alameda Watershed is dominated by non-native grasslands and oak woodlands, as well as oak savanna, shrublands, and riparian woodlands adjacent to channels (Table 2.1, 2.2). Non-native annual grassland, covering an estimated 46.3% of the watershed, likely occupies much of the area that historically supported native grassland or forbland habitat (Evett and Bartolome, 2013). Approximately 40.0% of the watershed is covered by oak savanna and woodland habitat, dominated by blue oak, coast live oak, and valley oak. Shrublands also cover much of the landscape, including chaparral (14.2%) and coastal scrub (5.4%). In addition to more common habitat types, the watershed supports serpentine grassland (0.7%), sycamore alluvial woodland (0.7%), white alder and willow riparian forest (0.4%, 0.4%), and other woodland types. Coastal or valley freshwater marshes occupy limited areas near waterways. This mosaic of vegetation (Fig. 2.1, 2.2) determines the large-scale patterns of carbon storage in the watershed and sets the stage for management activities focused on carbon sequestration.

Management activities may influence a landscape's carbon storage by altering the extents of grassland, shrubland, and woodland, or by changing the structure or carbon storage of existing ecosystem types. For sites in coastal California, ecosystem carbon storage tends to increase along a continuum from sites with herbaceous vegetation to shrublands to woodlands to forest. Larger trees, in denser stands, store more vegetation carbon per unit area than sparsely wooded sites or open grasslands with no woody cover (Silver et al., 2010; Zavaleta and Kettley, 2006). Belowground, the carbon stored in soils appears to follow this same trend. A synthesis study of rangeland soils in California found that soil carbon density (MT C/acre) was greater under a woody canopy than in open grassland (Silver et al., 2010), and a study in Marin County found that soil carbon stocks increased over time following shrub invasion of a grassland (Zavaleta and Kettley, 2006).

Without altering the distribution of grassland and woody vegetation, management choices affect the rates of carbon uptake and loss through photosynthesis, decomposition, and fire. In grasslands typical of the Alameda Watershed, soil carbon stocks are affected by the composition of herbaceous species, the physical treatment or

**Table 2.1. Vegetation composition of the Alameda Watershed, based on LANDFIRE data (as aggregated by Gonzalez et al. (2015)).** The more exhaustive vegetation classes correspond to the LANDFIRE-based classification system used in Gonzalez et al. (2015). The ecosystem types identified in the first column were used to quantify and summarize carbon storage across the watershed (see Chapter 3), and to characterize opportunities for carbon management activities (see Chapter 4). Because Jones and Stokes mapping (2003; Table 2.2) was used to reclassify a fraction of LANDFIRE-designated grassland, coastal scrub, and chaparral as oak savanna, the sum of acreages in this table differ slightly from those used in this study (e.g., Fig. 2.2, Table 3.3).

| Ecosystem type | Vegetation class from Gonzalez et al., (2015), crosswalked from LANDFIRE existing vegetation type | Acreage | % Area |
|---|---|---|---|
| Chaparral | California mesic chaparral | 2,550 | 6.6 |
| | Northern and central California dry-mesic chaparral | 1,511 | 3.9 |
| | Southern California dry-mesic chaparral | 1,071 | 2.8 |
| | Shrubland | 34 | 0.1 |
| Coastal scrub | Southern California coastal scrub | 2,242 | 5.8 |
| Grassland | Grassland | 13,278 | 34.6 |
| | Mediterranean California sparsely vegetated systems | 56 | 0.1 |
| Oak savanna | California lower montane blue oak-foothill pine woodland and savanna | 478 | 1.2 |
| | Deciduous sparse tree canopy | 19 | 0.05 |
| | Mixed evergreen-deciduous sparse tree canopy | 72 | 0.2 |
| | Southern California oak woodland and savanna | 2,206 | 5.8 |
| Oak woodland | California montane Jeffrey pine (ponderosa pine) woodland | 1 | 0.003 |
| | California montane woodland and chaparral | 1 | 0.003 |
| | Central and southern California mixed evergreen woodland | 8,122 | 21.2 |
| | Evergreen open tree canopy | 921 | 2.4 |
| | Mediterranean California dry-mesic mixed conifer forest and woodland | 0.21 | 0.0013 |
| | Mediterranean California mixed oak woodland | 476 | 1.2 |
| Riparian forest | California montane riparian systems | 657 | 1.7 |
| Open water, barren, or developed | Open water, barren, or developed | 4,667 | 12.2 |
| **Total** | **Total** | **38,363** | |

**Table 2.2. Vegetation composition of the Alameda Watershed, as mapped by Jones and Stokes (2003).** The total extent of Jones and Stokes 2003 mapping is less than the study area used in this analysis.

| Vegetation class | Acreage | % Area |
|---|---|---|
| Blue oak woodland | 1,364 | 3.7 |
| Central coast live oak riparian forest | 177 | 0.5 |
| Coast live oak riparian forest | 55 | 0.1 |
| Cultivated | 374 | 1 |
| Diablan sage scrub | 1,811 | 4.9 |
| Fresh marsh | 21 | 0.1 |
| Homes | 724 | 2 |
| Mixed evergreen forest/oak woodland | 9,612 | 26.1 |
| Natural pond | 1 | 0.003 |
| Nonnative grassland | 17,048 | 46.3 |
| Nurseries | 177 | 0.5 |
| Oak savanna | 1,194 | 3.2 |
| Quarry pond | 96 | 0.3 |
| Reservoir | 2,216 | 6 |
| Rock outcrop | 16 | 0.04 |
| Serpentine foothill pine-chaparral woodland | 72 | 0.2 |
| Serpentine grassland | 242 | 0.7 |
| Stock pond | 35 | 0.1 |
| Sycamore alluvial woodland | 275 | 0.7 |
| Valley oak woodland | 1,027 | 2.8 |
| White alder riparian forest | 136 | 0.4 |
| Willow riparian forest | 157 | 0.4 |
| Unmapped | 1 | 0.003 |
| **Total** | **36,831** | |

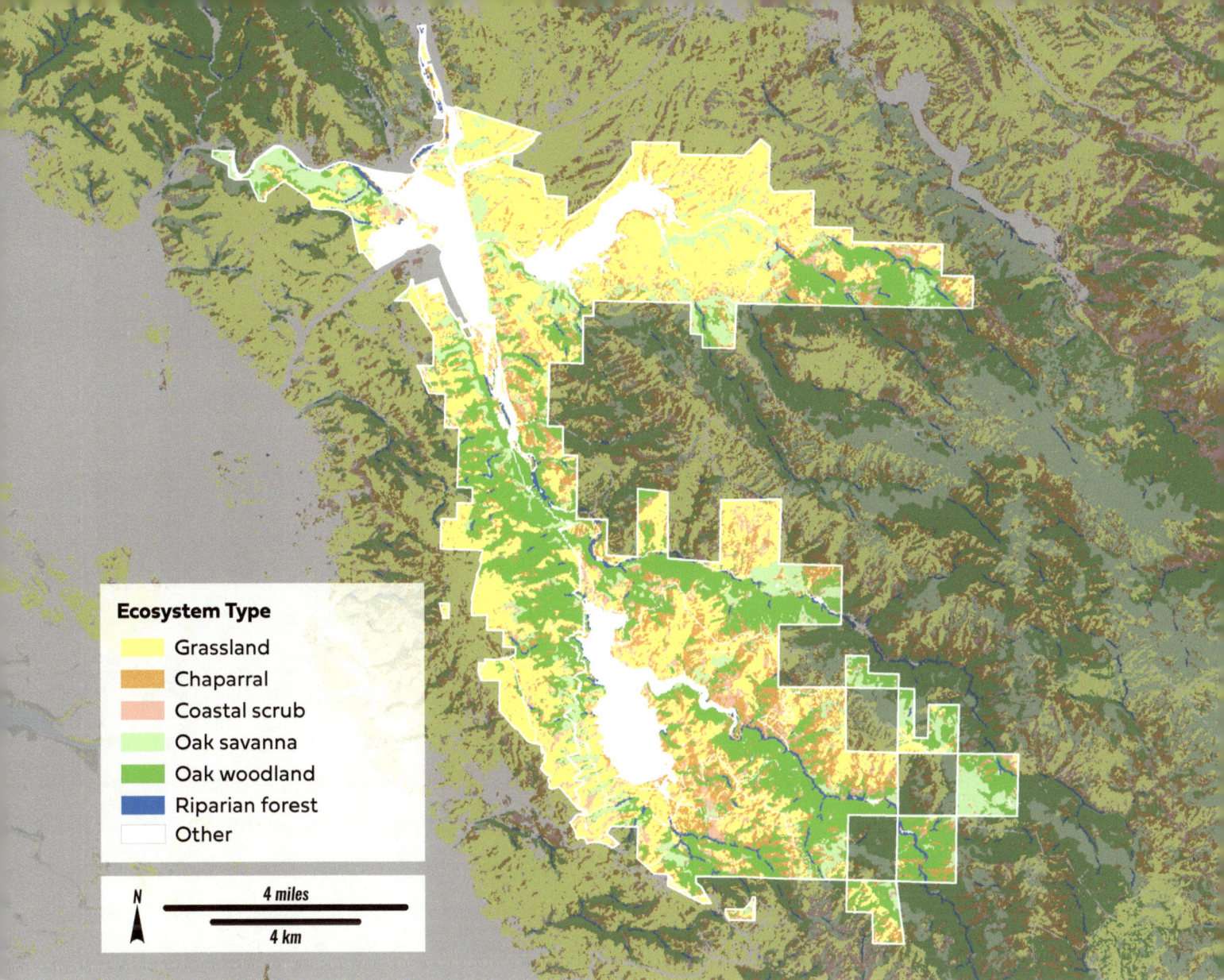

Figure 2.1. Distribution of generalized ecosystem types in the watershed. Ecosystem types were crosswalked from Gonzalez et al. (2015) vegetation classes (Table 2.1, based on 2010 LANDFIRE Expected Vegetation Type data). Because the composition of oak savanna is not well captured by the source imagery's 30 x 30 grid, Jones and Stokes (2003) mapping (Table 2.2) was used to define the extent of oak savanna.

**Ecosystem Type**

- Grassland
- Chaparral
- Coastal scrub
- Oak savanna
- Oak woodland
- Riparian forest
- Other

N

4 miles

4 km

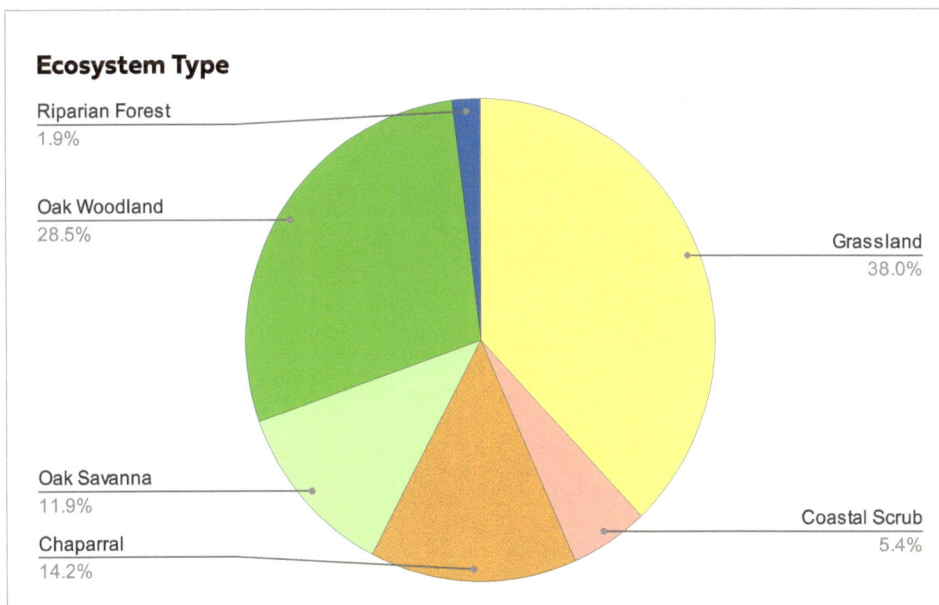

**Ecosystem Type**

Riparian Forest
1.9%

Oak Woodland
28.5%

Grassland
38.0%

Oak Savanna
11.9%

Chaparral
14.2%

Coastal Scrub
5.4%

Figure 2.2. Relative cover of dominant ecosystem types in the Alameda Watershed.

(Top) Serpentine outcrop. (Bottom) Blazing star flowers. Photographs courtesy of SFPUC.

disturbance of soil, and amendments to the soil that alter its fertility and decomposition properties. In wooded sites, managers affect carbon by changing stand structure and density, or influencing patterns of tree mortality and regrowth. Pathogen introductions, fire risk reduction treatments such as thinning and prescribed burns, and post-fire management practices all affect both short-term and long-term patterns of woodland carbon storage.

## HISTORY OF HUMAN DISTURBANCE

Human activities have altered carbon stocks on the Alameda Watershed since long before the watershed came under San Francisco Public Utilities Commission (SFPUC) management. Indeed, the history of the region is one of human influence, in which anthropogenic changes to the distribution of vegetation types and species composition have translated to carbon losses or gains. Indigenous management (including the controlled use of fire), timber harvest, livestock grazing, cultivation, introduction of non-native species, reservoir construction, construction of transportation corridors, fire suppression, contemporary watershed management and restoration, and other land management practices have all influenced the carbon balance of the watershed over the past two and a half centuries.

Prior to Euro-American colonization, the watershed and surrounding areas supported a mosaic of grassland, coastal scrub, chaparral, oak savanna and woodland, mixed evergreen forest, riparian forest and scrub, and seasonal wetland (Stanford et al., 2013).[2] Valley floor settings, such as Sunol Valley, generally supported grassland or oak savanna dominated by species such as valley oak (*Quercus lobata*) and blue oak (*Q. douglasii*). Riparian corridors along streams were characterized by sycamore alluvial woodland (within the alluvial reaches of Alameda Creek and other major channels) dominated by California sycamore (*Platanus racemosa*), mixed riparian forest, willow riparian scrub, and other riparian habitat types. While evidence for the pre-settlement vegetation cover in upland settings is sparse, the available data suggests that hillslopes were generally dominated by a mix of grassland and Diablan sage scrub or chaparral, with relatively little tree cover.

At the time of Spanish settlement the watershed was home to people of the Chochenyo Ohlone language group, whose ancestors had lived in the region for thousands of years. The native Ohlone inhabitants lived in a number of distinct local tribal groups; major groups within the watershed included the Causen in Sunol Valley and the Taunan along Upper Alameda Creek (Milliken, 1995; Milliken et al., 2010). The Ohlone used a wide variety of natural resources and carefully managed the landscape in a number of important ways, including through the use of fire. Spanish colonists displaced the Ohlone from their land, and many of the watershed's native inhabitants were forcibly assimilated at Mission San José. During the late 18th through mid-19th century, Spanish and Mexican settlers grazed livestock on rangelands throughout the region. Portions of the Alameda Watershed were included within grazing lands used by the mission, and later within Mexican land grants such as Valle de San Jose. Grazing continued under lease arrangements following acquisition of the watershed by the Spring Valley Water Company and later by SFPUC. Under the SFPUC's current rangeland management program, approximately 31,000 acres are leased for grazing within the watershed (as of 2020; SFPUC, 2017). The impact of grazing on carbon stocks in the watershed over time is not well understood: grazing can have variable effects on carbon sequestration depending on climate, grazing intensity and management, and other factors (Abdalla et al., 2018; Carey et al., 2020; Conant et al., 2017; McSherry and Ritchie, 2013).

In addition to grazing, dryland farming was widely practiced throughout the region during the 19th and early 20th centuries (Stanford et al., 2013). In some areas, farmers and homesteaders likely removed trees to clear land for cultivation, particularly in valley floor settings historically dominated by oak savanna such as La Costa Valley and Sunol Valley. Oaks in Sunol Valley, for example, had largely been cleared by the late 19th century, which presumably decreased carbon storage in this part of the watershed (Stanford et al., 2013). Another early impact of Euro-American settlement on the watershed was the spread of introduced plant species—particularly annual grasses and forbs—which displaced native grasslands and other vegetation types. By the early to mid-19th century, introduced species such as wild oat (*Avena* spp.) and filaree (*Erodium cicutarium*) had already displaced vast areas of native grassland and forbland in the Bay Area (Minnich, 2008). In other Bay Area sites, non-native annual grasslands have been found to have lower soil carbon storage than native perennial grasslands (Koteen et al., 2011).

Development of water supply infrastructure in the region commenced in the early 20th century with construction of Sunol Dam (completed 1900), Calaveras Dam

Early agriculture in Calaveras Valley, 1919. (D-92, image courtesy of the San Francisco Public Utilities Commission)

(completed 1925), and Turner Dam (completed 1965, creating San Antonio Reservoir; Hanson et al., 2005). The flooding of valley floor areas to create these reservoirs resulted in substantial losses of sycamore alluvial woodland, oak savanna, and other habitat types (Stanford et al., 2013). The net effect of artificial reservoirs on carbon storage and emissions depends on a number of factors, including climate, reservoir age, watershed geology, and others. In general, reservoirs in tropical regions tend to be carbon sources, while reservoirs in temperate regions tend to be carbon sinks (Phyoe and Wang, 2019).

At present, portions of the watershed are leased for grazing, tree nurseries and quarry operations to generate revenue. There is also a former golf course now under SFPUC management as grassland, which, having previously been amended with fertilizers, is a relevant site for soil carbon sampling to compare its soil carbon contents with the rest of the watershed. Another portion of the watershed is leased to the East Bay Regional Park District and has a trail network used for recreational hiking, cycling and horseback riding. §

Construction of Calaveras Dam, 1913. (C-312, image courtesy of the San Francisco Public Utilities Commission)

# 3 **CARBON STORAGE**
## IN THE ALAMEDA WATERSHED

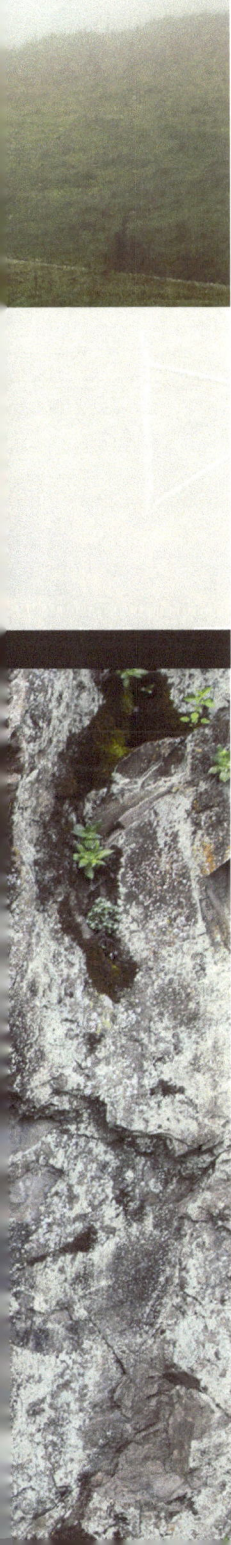

The ecosystems of the Alameda Watershed store carbon aboveground and belowground, in living vegetation, dead plant residues, and soil organic matter. In live vegetation, carbon is stored in woody and herbaceous biomass, both aboveground and in roots. Carbon is also stored in standing dead trees, downed dead wood, litter, and other accumulations of organic material on the soil surface. Finally, carbon stored in soil organic matter—plant and animal material in the process of decomposition—typically comprises a large percentage of total ecosystem carbon. The amount of carbon in each of these reservoirs, or *carbon pools*, varies across the watershed according to vegetation structure, soil properties, and the history of natural disturbances and human land use.

This chapter provides spatially explicit estimates of carbon storage across the Alameda Watershed. These estimates are based on a variety of data and models, which were used to map carbon stocks for a comprehensive set of carbon pools at the 30 m scale (Table 3.1), and provide a refinement to rough estimates developed for the watershed in 2016 from generalized carbon factors (Jones and Stokes, 2008; SF-PUC, 2016). Key advances of these updated numbers include the use of site-specific source data, a modeling approach that accounts for spatial variability in carbon storage within ecosystem types, and the inclusion of soil organic matter in carbon stock estimates. This analysis captures spatial variability in ecosystem carbon stocks, mean differences in carbon storage among ecosystem types, and the contributions of each carbon pool to total carbon storage.

The baseline carbon stock estimates provided in this chapter are based on satellite imagery and vegetation mapping from 2010 and 2014. In 2020, the SCU Lightning Complex fires burned nearly 30% of the watershed's area, reducing carbon storage in the footprint of the fire as grasses, shrubs, and other woodland fuels were consumed. This fire—the fourth largest recorded in California to date—provided an opportunity to quantify the immediate effects of a large wildfire event on ecosystem carbon storage. By coupling pre-fire 2010 carbon stock estimates with modeled carbon losses due to the SCU Lightning Complex, this assessment provides before and after snapshots that track short-term changes in carbon stocks and set a new, post-fire, baseline against which ecosystem recovery could be monitored. Together, this information quantifies the watershed's function as a carbon reservoir and sets the stage to consider carbon management in the context of increasing wildfire risk (Goss et al., 2020).

## ALAMEDA WATERSHED CARBON STORAGE

### Vegetation carbon

Carbon storage in vegetation was quantified with models and data sources (Table 3.1) specific to each carbon pool and ecosystem type described in Chapter 2 (see Fig. 2.1). In areas with a woody canopy—classified in this analysis as riparian forest, oak woodland, oak savanna, chaparral, or coastal scrub—aboveground biomass carbon estimates were based on gridded data products available for 2010 and 2014 through the US Forest Service and Department of the Interior's LANDFIRE program. Carbon in standing trees and shrubs was quantified according to Gonzalez et al. (2015), using LANDFIRE Existing Vegetation Type, Forest Canopy Height, and Forest Canopy Cover layers. Equations used in this analyses were developed for California forests and shrublands by Gonzalez et al. (2015) from US Forest Service Forest Inventory Analysis plots and the published literature. Carbon in understory herbaceous vegetation, dead wood, litter, and duff was derived from Fuel Characteristics Classification System (FCCS) data (Prichard et al., 2019; Reeves et al., 2009; Riccardi et al., 2007). FCCS fuelbeds provide gridded biomass estimates at 30 m resolution for a range of carbon pools, and provide reasonable estimates of dead biomass stocks for the purpose of carbon accounting (Saah et al., 2016). Vegetation biomass values were converted to units of carbon using a biomass carbon fraction of 0.47 (IPCC, 2006).

In areas classified as grassland, aboveground carbon in herbaceous vegetation was estimated from the watershed-wide average of 2018 biomass measurements from the SFPUC residual dry matter (RDM) monitoring program (ACRCD and LD Ford, 2018a). (As discussed in the technical appendix, RDM plots were not designed to provide a random sample across the watershed or individual grazing leases, so the average RDM value should be viewed as an approximation.) In grassland sites, carbon in shrubs, dead wood, litter, and duff was based on FCCS fuelbeds. For both woody and herbaceous sites, root-to-shoot ratios from the published literature were used to estimate carbon storage in belowground vegetation.

Carbon storage in vegetation ranged from 1.2 MT C/acre in low-biomass grassland to as high as 125 MT C/acre in riparian forest sites with the greatest canopy cover and height. In general, modeled vegetation carbon storage was greatest in ecosystems that have dense woody vegetation such as riparian forest or oak woodlands, so the map of vegetation carbon (Fig. 3.1) resembles the distribution of the watershed's woodlands and riparian forest (Fig. 2.1). These high-carbon areas are concentrated on canyon bottoms along riparian corridors and north-facing slopes where tree cover is high. More common on south-facing slopes, shrublands (chaparral and coastal scrub) were found to store low to moderate carbon ranging from 8.2 MT C/acre in sparse coastal scrub sites to 31 MT C/acre in the most dense chaparral stands.

Carbon storage in dead wood, litter and duff varied widely among ecosystem types. Although this is an important carbon pool, accounting for as much as 17 MT C/acre in riparian forest, or 65% of total vegetation carbon in coastal scrub sites, it is very difficult to measure in field studies and contributes to uncertainty in carbon inventories (Saah et al., 2016). Accordingly the spatial error ranges reported in Table 3.2 may not fully capture the uncertainty in dead wood carbon stocks. Similarly, spatial standard deviations for standing trees and shrubs do not capture model uncertainty associated with LANDFIRE layers and model equations (Gonzalez et al., 2015).

**Table 3.1: Carbon pools and carbon quantification methods for grassland and woody-dominated sites.** Woody-dominated sites include those classified as coastal scrub, chaparral, savanna, woodland, or riparian forest. A complete description of carbon quantification methods is provided in the online technical appendix (https://www.sfei.org/projects/alameda-watershed-carbon).

| Carbon pool | Description | Source of carbon estimates: grassland sites | Source of carbon estimates: woody-dominated sites |
|---|---|---|---|
| Trees | Live and dead standing trees | – | Modeled from LANDFIRE existing vegetation type, canopy cover, and canopy height (Gonzalez et al., 2015) |
| Shrubs | Low woody plants with multiple stems | FCCS values | Modeled from LANDFIRE existing vegetation type, canopy cover, and canopy height (Gonzalez et al., 2015) |
| Herbaceous vegetation | Live and dead herbaceous vegetation | RDM measurements (simple average of ~100 plots across the watershed grazing leases) | FCCS values |
| Roots | Live and dead roots | Root:shoot ratios applied to RDM values | Root:shoot ratios applied to tree and shrub values |
| Dead wood, litter, and duff | Downed and dead woody material, litter and duff (O horizons), lichen, moss, cryptogams, basal accumulations around tree trunks, and squirrel middens | FCCS values | FCCS values |
| Soil | Mineral soil to 1 m depth | Synthesis of literature values and Alameda Watershed measurements | Synthesis of literature values and Alameda Watershed measurements |

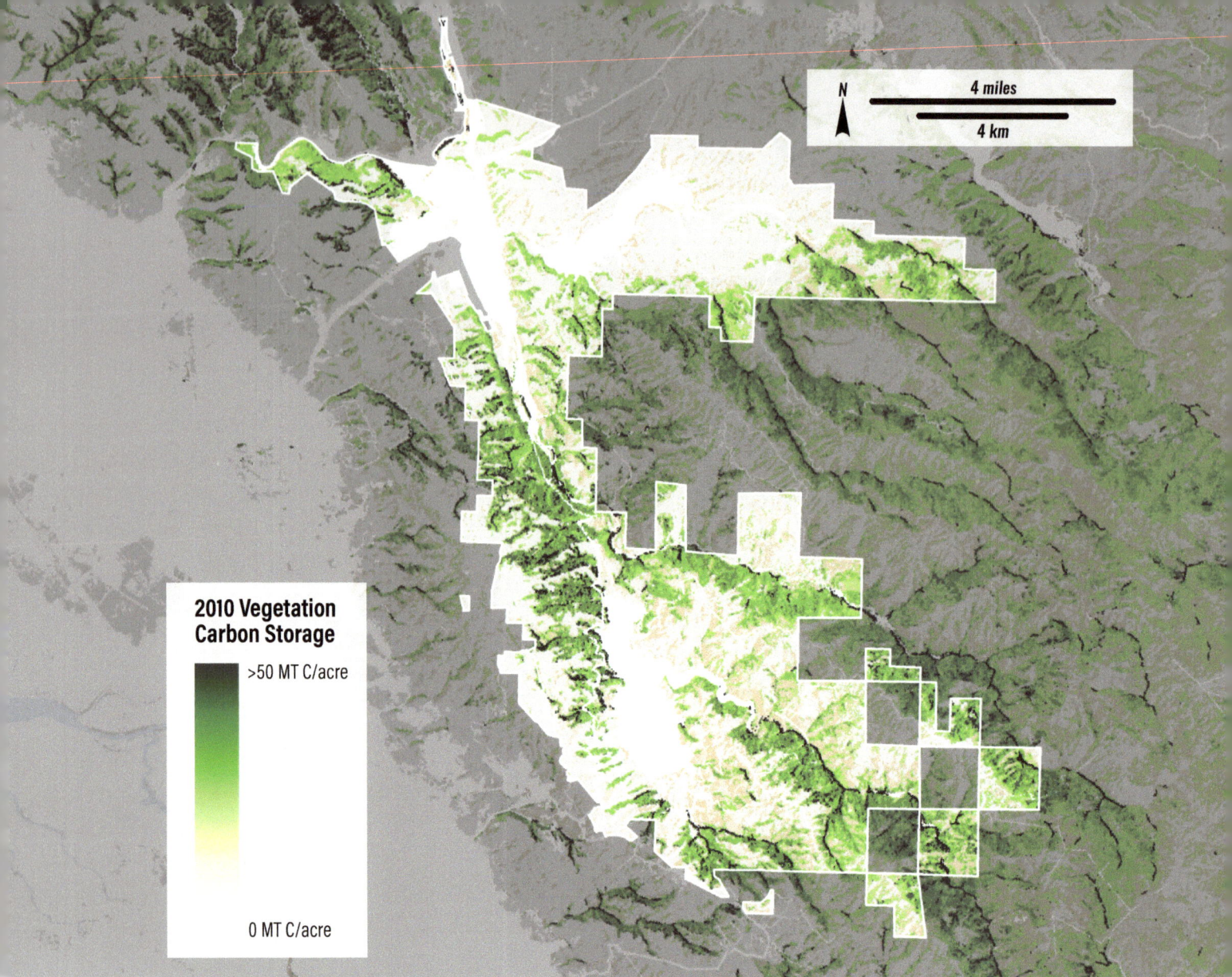

**2010 Vegetation Carbon Storage**

>50 MT C/acre

0 MT C/acre

**Figure 3.1. Carbon storage in vegetation.** Vegetation carbon includes trees, shrubs, herbaceous vegetation, roots, dead wood, litter, and duff.

**Table 3.2. Carbon stocks in the Alameda Watershed's vegetation carbon pools.** Carbon stocks were estimated according to the approaches in Table 3.1. Error ranges indicate the standard deviation in carbon storage across grid cells of a given ecosystem type, with the exception of herbaceous vegetation in grassland, for which the error range is the standard deviation across RDM measurements.

| Ecosystem type | Trees (MT C/acre ± 1σ) | Shrubs (MT C/acre ± 1σ) | Herbaceous vegetation (MT C/acre ± 1σ) | Dead wood, litter, and duff (MT C/acre ± 1σ) | Roots (MT C/acre ± 1σ) |
|---|---|---|---|---|---|
| Grassland | 0 | 0.0014 ± 0.02 | 0.6 ± 0.26 | 0.036 ± 0.13 | 0.55 ± 0.01 |
| Coastal scrub | 0 | 1.4 ± 0 | 0.015 ± 0.1 | 4.0 ± 0.41 | 0.74 ± 0.1 |
| Chaparral | 0 | 6.8 ± 2.2 | 0.0064 ± 0.1 | 3.1 ± 0.26 | 3.6 ± 01.2 |
| Oak savanna | 10.6 ± 6.3 | 1.5 ± 1.4 | 0.13 ± 0.1 | 3.6 ± 2 | 5.6 ± 2.9 |
| Oak woodland | 18.5 ± 7.3 | 1.7 ± 1 | 0.16 ± 0.1 | 3.9 ± 0.6 | 9.3 ± 3.4 |
| Riparian forest | 30.1 ± 7.9 | 2.1 ± 0.2 | 0.099 ± 0.007 | 17.5 ± 2.5 | 14.7 ± 3.5 |

## Soil carbon

Carbon storage in the watershed's soils was estimated from a synthesis of published measurements from California as well as soil samples collected in 2021 from the Alameda Watershed. The synthesis included a total of 65 soil profiles from 23 grassland and woodland sites outside the Alameda Watershed, plus 16 soil profiles collected from four paired grassland and woodland transects south of the San Antonio Reservoir. The depth to which soils were sampled varied widely across studies in the synthesis, ranging from 2 cm to 4.6 m below the mineral soil surface. Carbon concentrations are typically highest in shallow soils, but substantial carbon resides at depth as well (Sulman et al., 2020), so the depth of sampling strongly influences the total reported soil carbon stock. This makes it critical to standardize carbon stock measurements to a common sampling depth before comparing or summarizing across studies or sites. To address this issue, the method described in Silver et al. (2010) was used to model the carbon stock to a standard depth of 1 m for each soil profile in the synthesis. For each soil profile, cumulative carbon storage was calculated at each depth that was sampled (e.g., 17 MT C/acre in the top 25 cm, 26 MT C/acre in the top 50 cm, and 30 MT C/acre in the top 50 cm). Cumulative carbon storage was plotted against depth for the entire dataset (Fig. 3.2), and the fit of this regression was used to estimate each soil profile's carbon stock to a standard depth of 1 m. (Detailed methods information is provided in the online technical appendix.)

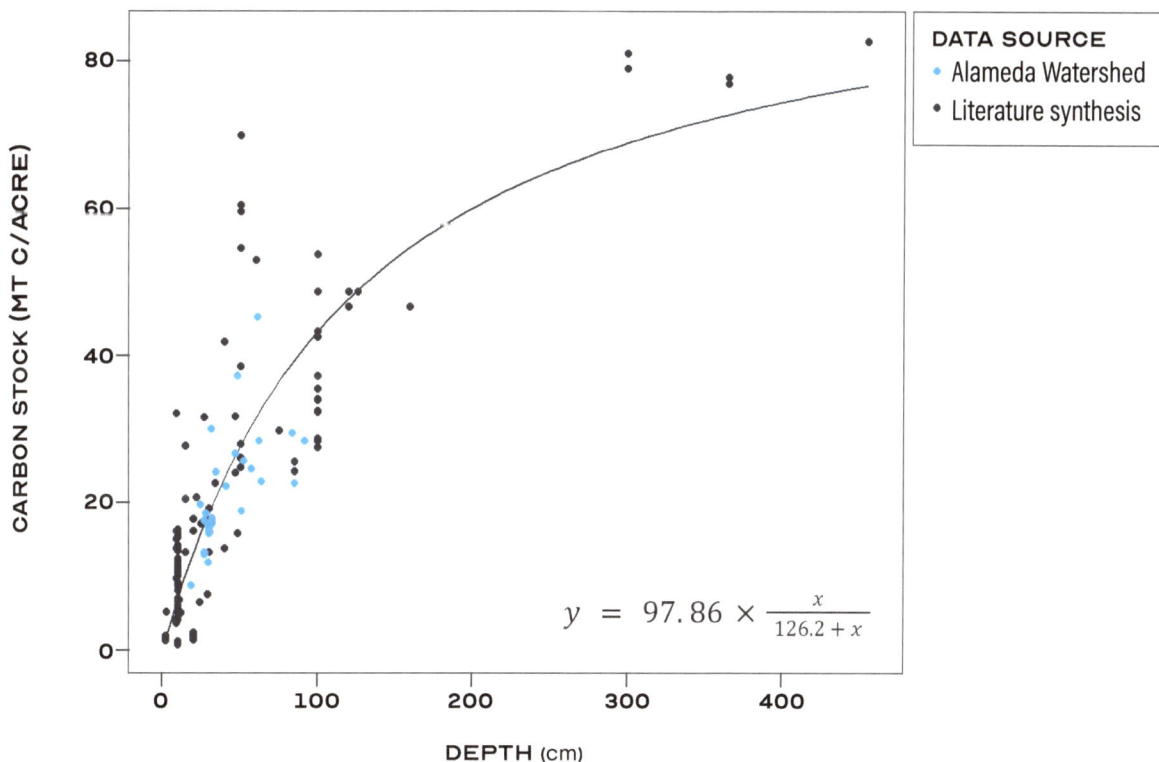

$$y = 97.86 \times \frac{x}{126.2 + x}$$

**Figure 3.2. Relationship between soil sampling depth and cumulative carbon storage** for measurements included in the soil carbon synthesis. Black points indicate values from the literature, from sites outside the Alameda Watershed. Blue points represent measurements from grassland and woodland sites in the Alameda Watershed, south of the San Antonio Reservoir. The calculated regression equation was used to estimate carbon storage to 1 m depth for each measurement.

Across soil profiles included in the synthesis, modeled carbon stocks to 1 m depth ranged widely, with most data falling between 4.3 and 120 MT C/acre, and a single high value of 210 MT C/acre (Fig. 3.3). Average soil carbon stocks were significantly higher in sites with a woody canopy (avg = 64.06 ± 36.08 MT C/acre) than in open grassland (avg = 46.0 ± 26.68 MT C/acre; Fig. 3.3). This pattern has been documented before in a California-wide synthesis (Silver et al., 2010), and in observational studies at individual sites (e.g., Camping, 2002; Dahlgren et al., 1997; Herman et al., 2003; Zavaleta and Kettley, 2006).

Results of the soil carbon data synthesis were applied to the Alameda Watershed according to mapped ecosystem type. The grassland average of 46.01 MT C/acre was applied to areas classified as grassland, and the woody canopy average of 64.06 MT C/acre was applied to areas classified as coastal scrub, chaparral, oak savanna, oak woodland, or riparian forest. This simplified approach assumes that the watershed's soils follow general patterns seen elsewhere in the state, and offers first-order, spatially averaged estimates. Actual carbon storage varies widely among and within sites, as illustrated by the broad range of 1 m carbon storage values shown in Fig. 3.3. Soil carbon storage depends on site-specific properties such as vegetation, hydrology, geomorphology, soil characteristics, and microclimate. For example, a study in Marin County found that across a topographic sequence, soil carbon storage increased ~3-fold from the eroding summit to the depositional plain (Berhe et al., 2012). The first-order estimates provided in this analysis are not able to account for this type of spatial variation. More precise soil carbon mapping across the Alameda Watershed would require extensive on-site sampling.

Carbon stock measurements from Alameda Watershed samples—which ranged from 25 to 61 MT C/acre—fell within the general range of other studies included in the synthesis (Fig. 3.2). This comparison suggests that synthesis results offer a reasonable approximation of average watershed-wide soil carbon storage. Grassland sites in particular did not differ between Alameda Watershed samples and the rest of the synthesis dataset (Fig. 3.3). Samples collected under trees on the watershed had marginally lower carbon storage than other values in the synthesis (not statistically significant; Fig. 3.3), but these samples were limited to only six soil profiles from two transects, five from a ridgeline and one from a valley floor adjacent to a seasonal creek. Values from this synthesis were also similar to soil carbon stocks from nearby sites reported in the International Soil Carbon Network (ISCN) Database (43–57 MT C/acre, depending on whether a high value is considered an outlier; https://iscn.fluxdata.org/data/), which includes user-uploaded data as well as carbon stocks reported by the National Resource Conservation Service (NRCS). (No values from within the study boundary were available through the ISCN database, so nearby sites were used for comparison.)

## Total ecosystem carbon

Soil and vegetation carbon estimates were combined to produce a map of ecosystem carbon stocks across the Alameda Watershed (Fig. 3.4). Watershed-wide, the site was found to store a total of 540,000 MT C in vegetation and 1.9 MMT C in soil, for a total of 2.5 MMT of carbon. Fig. 3.5 and Table 3.3 provide a summary of carbon stocks for the six primary ecosystem types: grassland, coastal scrub, chaparral, oak savanna, oak woodland, and riparian forest. (Developed areas and aquatic habitats were excluded from this analysis.) On average, the

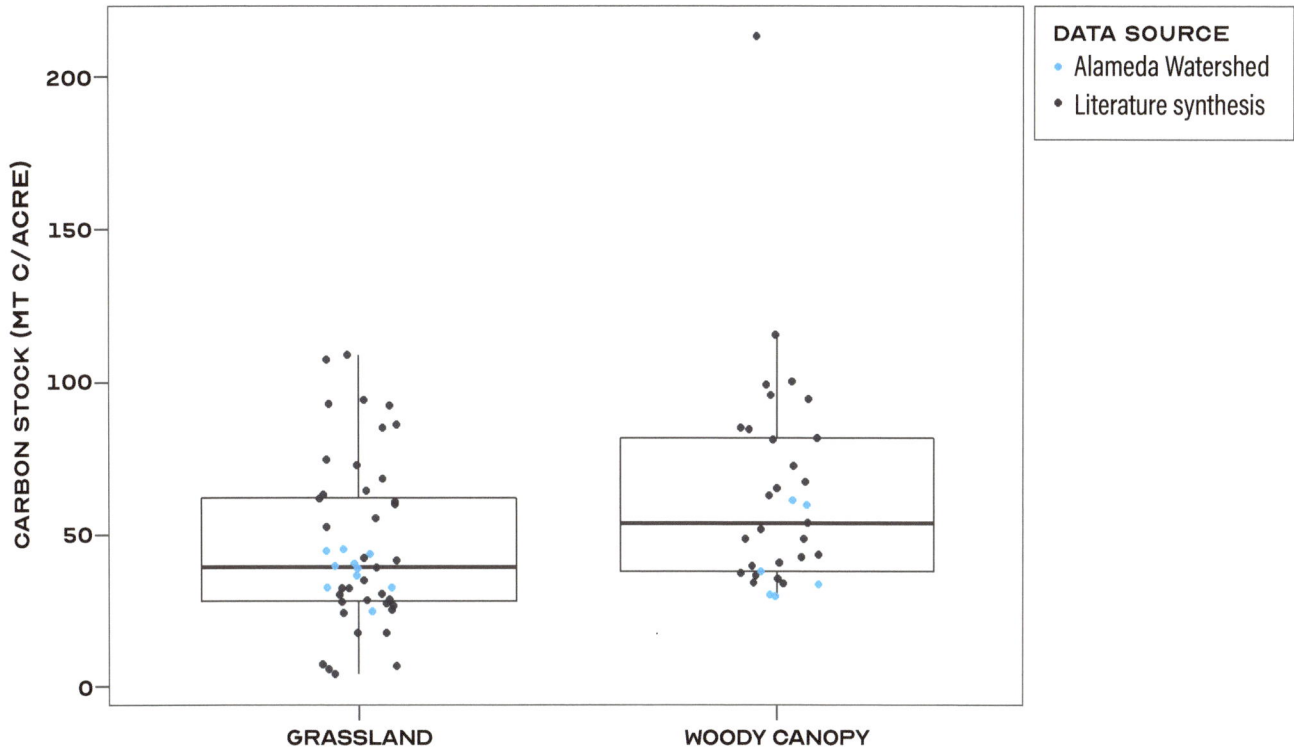

**Figure 3.3. Modeled carbon stocks to 1 m depth for 81 soil profiles across California.** Black points indicate values from the literature, from sites outside the Alameda Watershed. Blue points represent measurements from grassland and woodland sites in the Alameda Watershed, south of the San Antonio Reservoir. Boxes indicate the median and upper and lower quartiles across all data for grassland or woody-canopy sites, with whiskers extending to 1.5x the interquartile range. The average across sites with a woody canopy (64.06 MT C/acre) was significantly higher (p = 0.0115) than the average across grassland sites (46.01 MT C/acre). Alameda Watershed values did not differ significantly from other values in the synthesis at p = 0.05.

watershed's grasslands were found to store the least carbon per acre of the six ecosystem types, with an average carbon density of 47.2 MT C/acre (Table 3.3). Given their large spatial footprint, however, grassland carbon accounts for a large percentage of the watershed's total carbon (22%, or 601,700 MT C). In contrast, riparian forest had the highest per-acre carbon storage (128.5 MT C/acre on average) but the lowest watershed-wide total given its limited spatial extent (83,900 MT C over 653 acres). Carbon storage in the other ecosystem types ranged widely, both among ecosystem types and spatially within a given ecosystem type (Table 3.3). Of these, oak woodland stands out for having relatively high per-acre carbon storage and wide spatial coverage. With an average carbon density of 97.6 MT C/acre and a total extent of ~9,600 acres, non-riparian woodland sites were found to store ~930,000 MT C, or 38% of the watershed's total carbon.

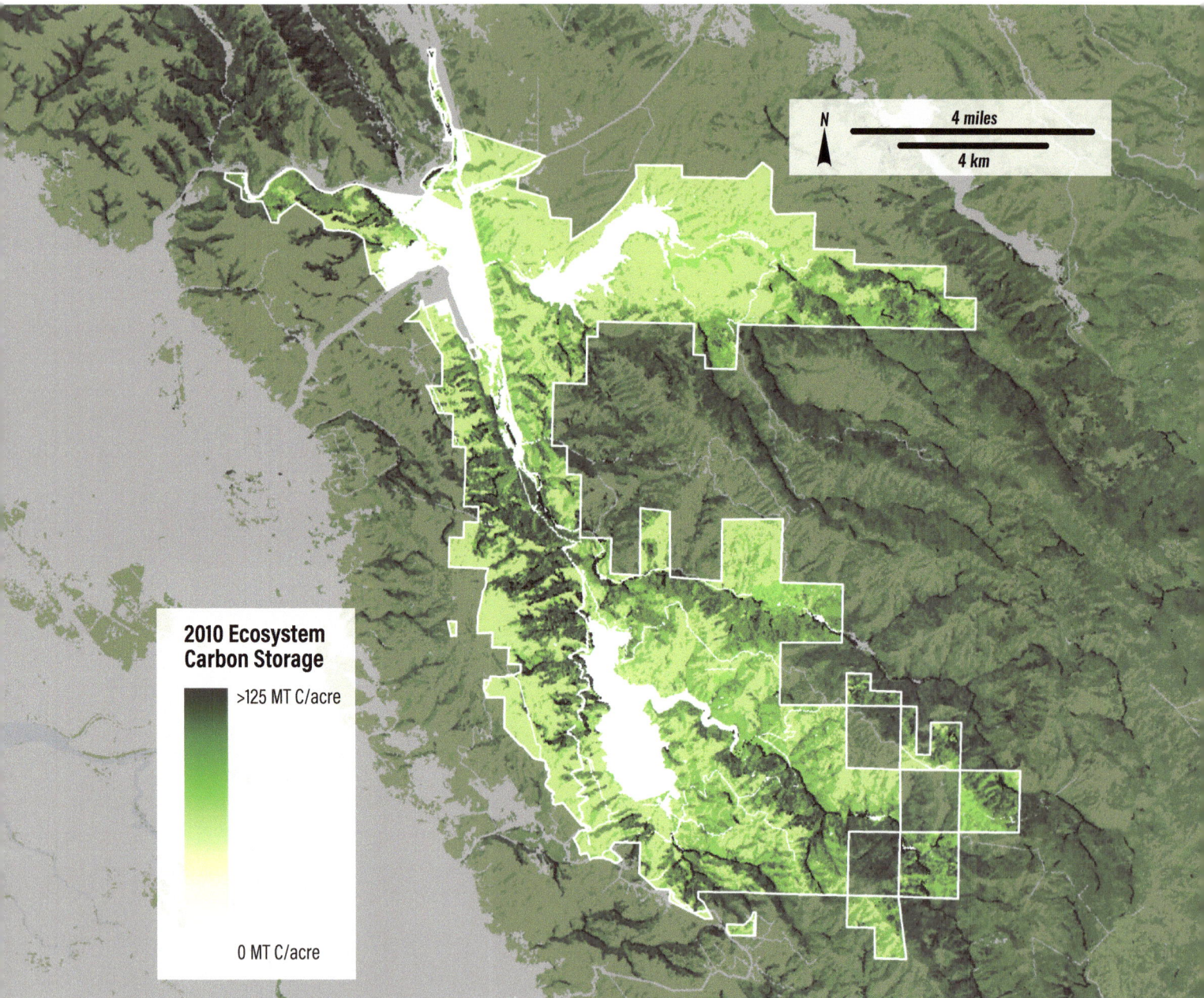

**Figure 3.4. Ecosystem carbon storage in the Alameda Watershed.** Carbon storage includes both vegetation and soils for sites not classified as water, barren, or developed.

**Table 3.3. Carbon storage within Alameda Watershed ecosystem types, per acre (a) and watershed-wide (b).**
In (a), error values for vegetation carbon represent spatial variation in per-acre carbon stocks across 30 m pixels for a given ecosystem type. Soil carbon error values represent the standard error from the data synthesis. In (b), values in parentheses indicate the percentage of the total. Additional land cover types not evaluated in this assessment include water bodies and developed areas.

**Table 3.3 (a)**

| Ecosystem type | Vegetation Carbon Storage (MT C/acre ± 1 σ) | Soil Carbon Storage (MT C/acre ± 1 SE) |
|---|---|---|
| Grassland | 1.2 ± 0.14 | 1.2 ± 0.14 |
| Coastal scrub | 6.1 ± 0.25 | 64.1 ± 6.3 |
| Chaparral | 13.5 ± 3.5 | 64.1 ± 6.3 |
| Oak savanna | 21.4 ± 10.8 | 64.1 ± 6.3 |
| Oak woodland | 33.5 ± 10.8 | 64.1 ± 6.3 |
| Riparian forest | 64.4 ± 11.8 | 64.1 ± 6.3 |

**Table 3.3 (b)**

| Ecosystem type | Area (acres and percent of total) | Vegetation Carbon Storage (MT C) | Soil carbon storage (MT C) | Total ecosystem carbon (MT C and percent of total) |
|---|---|---|---|---|
| Grassland | 12,744 (38%) | 15,400 | 586,000 | 602,000 (24%) |
| Coastal scrub | 1,815 (5%) | 11,100 | 116,000 | 127,000 (5%) |
| Chaparral | 4,777 (14%) | 64,500 | 306,000 | 371,000 (15%) |
| Oak savanna | 3,988 (12%) | 85,400 | 255,000 | 341,000 (14%) |
| Oak woodland | 9,557 (29%) | 320,500 | 612,000 | 933,000 (38%) |
| Riparian forest | 653 (2%) | 42,100 | 42,000 | 84,000 (3%) |
| Total | 33,534 | 539,100 | 1,918,000 | 2,457,000 |

RIPARIAN FOREST

OAK WOODLAND

OAK SAVANNA

CHAPARRAL

COASTAL SCRUB

GRASSLAND

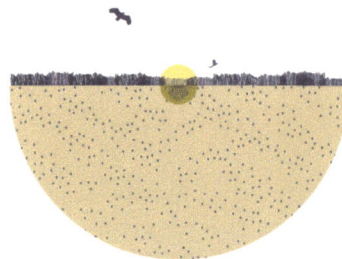

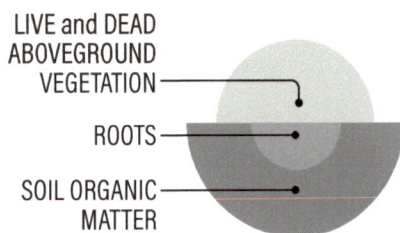

LIVE and DEAD ABOVEGROUND VEGETATION

ROOTS

SOIL ORGANIC MATTER

**Figure 3.5. Per-acre carbon storage for ecosystem types in the Alameda Watershed.** The size of each semi-circle represents the relative amount of carbon per acre stored in aboveground live and dead vegetation, roots, and soil organic matter for each of the six major ecosystem types. The relative size of each carbon pool corresponds to values in Tables 3.2 and 3.3a.

A comparison of aboveground and belowground carbon pools demonstrates the importance of soil as a carbon reservoir (Fig. 3.5). Roughly half the carbon in riparian forest sites resides in soil organic matter. In the watershed's grasslands, soil carbon accounts for an estimated 97% of total ecosystem carbon, with only ~3% in aboveground vegetation and roots.

Relative to other ecosystems across the globe, the Alameda Watershed has moderate carbon storage. This aligns with expectations for Mediterranean-type ecosystems presented in Chapter 2. Averaged across this study's full 33,534 acres, the watershed was found to store ~16 MT C/acre in vegetation, only slightly higher than the global average from Mediterranean-type ecosystems from the Intergovernmental Panel on Climate Change (IPCC) tier 1 values (13 MT C/acre; Gibbs and Ruesch, 2008). The watershed's estimated soil carbon storage, 57 MT C/acre on average, is slightly higher than most other Mediterranean-type sites, where the top meter of soil commonly stores between 25 and 50 MT C/acre (Batjes, 2016). This may reflect uncertainty in soil carbon quantification—at both the watershed and global scales—or it may indicate that California has relatively high soil carbon storage compared with other regions with a similar climate. The Alameda Watershed's mosaic of grassland, shrubland, and woodland ecosystems has higher per-acre carbon storage than the state's extensive grasslands, but considerably lower carbon storage than other ecosystems in California. For example, field carbon inventories have reported aboveground carbon stocks (excluding roots) of 130 MT C/acre from an old-growth coastal redwood site, 110 MT C/acre from Sierra Nevada oak, Douglas fir, and mixed-elevation conifer forests, and 200 MT C/acre from a red fir forest (Gonzalez et al., 2010). Seen in this light, the ecosystems of the Alameda Watershed represent an important carbon reservoir within the Bay Area, but a site whose overall carbon storage is limited by the native vegetation mosaic.

## 2020 WILDFIRE CARBON LOSSES

The carbon storage estimates provided above are based on LANDFIRE and FCCS data products from 2010 and 2014. In the 2020 fire season, the SCU Lightning Complex burned 10,370 acres across the watershed, largely concentrated in the southeastern and central watershed (Fig. 3.6). This event drove a net loss of carbon from the site, converting carbon stored in wildland fuels to $CO_2$, methane, and other climate pollutants.

The First Order Fire Effects Model (FOFEM) was used to estimate short-term carbon losses during the SCU Complex Fires. FOFEM was produced by the US Forest Service to predict the immediate effects of wildfire on vegetation and fuels (https://www.firelab.org/project/fofem-fire-effects-model), and is the model used by the California Air Resources Board (CARB) to track fire-related greenhouse gas (GHG) emissions (CARB, 2022). FOFEM simulations predict biomass consumption and combustion-related emissions, according to input values for the vegetation type, fuel structure, moisture content, and burn severity. FOFEM was run for each cell in a 30 m grid across the burned extent in the watershed using input values from 2014 FCCS fuelbeds, weather station data, and Landsat imagery.

The distribution of fuels used with FOFEM was set to match the biomass distribution from the watershed-wide carbon storage assessment. This fuel structure made it possible to directly relate wildfire carbon loss estimates to the carbon stocks mapped in Fig. 3.4. FCCS fuelbeds were used for litter, 1-hour, 10-hour, 100-hour, and 1000-hour fuels. Herbaceous vegetation and shrub biomass were set to equal the biomass values used in the watershed-wide carbon assessment, based on RDM measurements, FCCS fuelbeds, and/or LANDFIRE vegetation type, canopy cover,

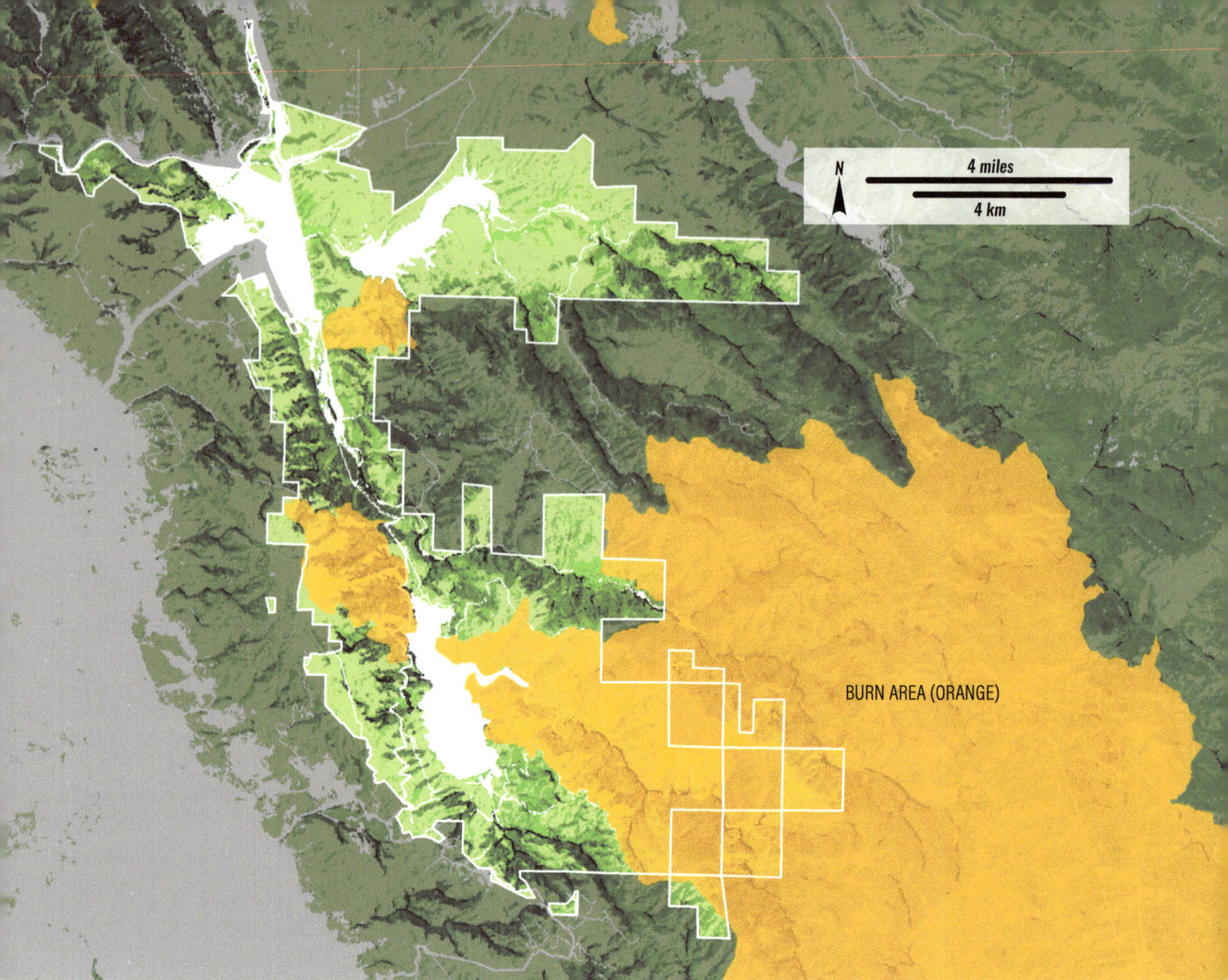

**Figure 3.6. Area burned in the 2020 SCU Lightning Complex Fires.**

and canopy height data (ACRCD and LD Ford, 2018a; Gonzalez et al., 2015; Riccardi et al., 2007). Foliage and branch densities were derived from LANDFIRE-based tree biomass values (Gonzalez et al., 2015) and allometric relationships (Brown et al., 2004; Chojnacky et al., 2014). Fuel moisture was based on averages reported by nearby Remote Automatic Weather Stations (RAWS) for the duration of the fire, and percentage crown burn was estimated from the relativized difference in normalized burn ratio (RdNBR), a burn severity metric derived from LANDSAT imagery (Miller et al., 2009; Miller and Thode, 2007), and an empirical relationship reported in Lydersen et al. (2016). More detailed descriptions of model inputs can be found in the technical appendix.

Results of the FOFEM analysis indicate that an estimated 33,100 MT C was lost from the Alameda Watershed during the SCU Lightning Complex fires (Table 3.4, Fig. 3.7). For grassland, coastal scrub, and chaparral sites, the model predicted that most vegetation carbon was lost in the fire, whereas carbon stored in trees—in oak savanna, oak woodland, and riparian forest sites—was largely retained on-site (Table 3.4). Accordingly, modeled per-acre carbon losses were generally highest from sites classified as chaparral (Fig. 2.1, 3.7), an ecosystem type with relatively high

Landscapes in the Alameda Watershed following the SCU Lightning Complex fires, photographs courtesy of SFPUC.

biomass carbon storage and high flammability, particularly in sites that have not recently burned (Keeley et al., 2011; Schwilk, 2003).

The majority of carbon consumed by wildfire is released to the atmosphere as $CO_2$. In addition, fire releases methane and other pollutants that can affect air quality and the climate (Clinton et al., 2006; Urbanski, 2009). Black carbon, for example, is a product of incomplete combustion that is released as an aerosol when biomass is burned. Although it resides in the atmosphere for a relatively short time, black carbon is understood to be one of the most important anthropogenic pollutants for both the climate and human health (Bond et al., 2013).

Over the span of just ~two months, the Alameda Watershed lost an estimated 8% of its vegetation carbon stock to fire. This rapid and large change, though likely short-lived, highlights the potential for wildfire to alter the distribution of stocks across California's natural and working lands. Carbon sequestered in the state's forests, woodlands, and shrublands is vulnerable to wildfire and other ecological disturbances. The long-term carbon balance of these systems depends on the frequency of fire relative to regeneration rates and whether changing fire regimes alter vegetation communities, for example if woodlands are replaced by grasslands or shrublands (Huntsinger and Bartolome, 1992; Lenihan et al., 2008). Statewide carbon inventories and future scenario modeling suggest

Table 3.4. Modeled change in carbon stocks due to the SCU Lightning Complex fires.

| Ecosystem type | Total area of ecosystem type (acres) | Area of ecosystem type burned (acres and % of total area) | Pre-fire aboveground carbon (MT C) | Post-fire aboveground carbon (MT C) | Carbon lost in fire (MT C and % of pre-fire aboveground carbon) |
|---|---|---|---|---|---|
| Grassland | 12,744 | 2,540 (20%) | 8,300 | 6,700 | 1,600 (19.2%) |
| Coastal scrub | 1,815 | 468 (26%) | 9,700 | 7,500 | 2,200 (22.6%) |
| Chaparral | 4,777 | 1,770 (37%) | 47,200 | 34,300 | 12,900 (27.4%) |
| Savanna | 3,988 | 1,198 (30%) | 63,200 | 59,900 | 3,300 (5.3%) |
| Woodland | 9,557 | 4,186 (44%) | 231,800 | 221,000 | 10,800 (4.7%) |
| Riparian forest | 653 | 208 (32%) | 32,500 | 30,300 | 2,200 (6.9%) |
| All | 33,534.5 | 10,370 (31%) | 392,800 | 359,700 | 33,100 (8.4%) |

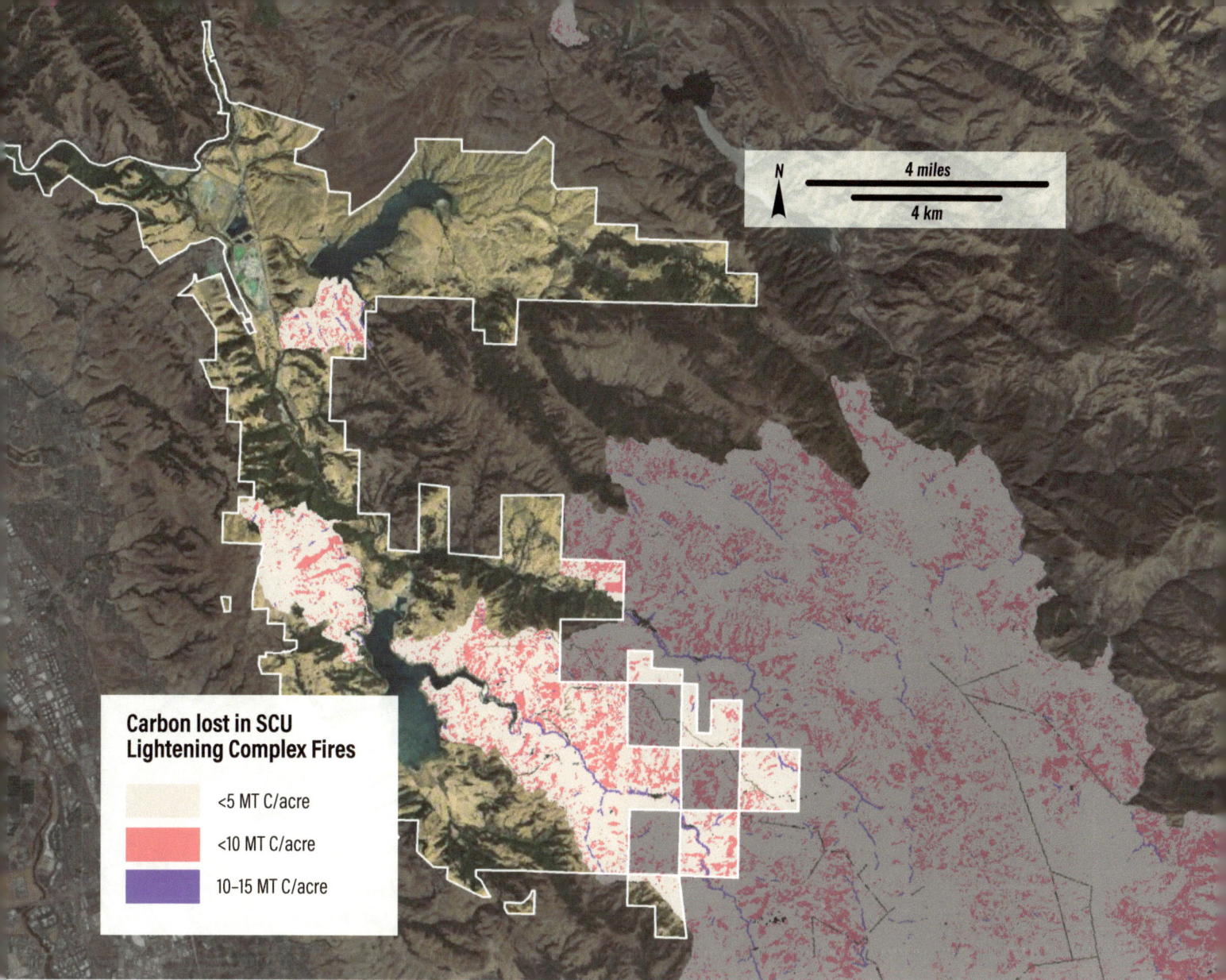

**Figure 3.7. Estimated carbon losses during the 2020 SCU Lightning Complex Fires.**

that California's wildlands are losing carbon due to wildfire, and that continued losses are expected even with increased fuel management (CARB, 2022b; Gonzalez et al., 2015). In the case of the 2020 fire, much of the carbon lost from the Alameda Watershed is likely to be re-captured as vegetation regenerates in the years following the fire, given the generally low burn severity in wooded sites and general resilience of shrubland and chaparral to fire (Keeley et al., 2011). The actual timing of regeneration and successional trajectories, however, will depend on specific species, burn severity, grazing by livestock, post-fire precipitation, and other variables that are challenging to predict (Keeley et al., 2005, 2008; Moreno and Oechel, 1991, 1992). In general, herbaceous vegetation regrows quickly post-fire, whereas vegetation cover and biomass in shrublands can take 10 to 15 years to recover to pre-fire levels (Hope et al., 2007; Keeley et al., 2011; Kinoshita and Hogue, 2011; McMichael et al., 2004). The long-term effects of fire on the watershed's carbon stocks will be borne out over the coming decades. §

Sunset, Alameda Watershed, courtesy of SFPUC.

# 4 MANAGING CARBON
## IN THE ALAMEDA WATERSHED

A number of strategies exist that land managers can implement to actively manage ecosystems for carbon sequestration. These approaches, known collectively as *carbon farming*, seek to increase ecosystem carbon stocks by increasing carbon inputs and/or decreasing carbon outputs from vegetation and soils. As managers look to ecosystems as a tool for climate change mitigation, the choice of whether, and how, to manage a given site for carbon should depend on a number of factors. These factors include not only the potential for a given strategy to provide carbon and greenhouse benefits, but also feasibility and cost, the long-term durability of carbon benefits, and potential impacts to other ecosystem functions and management priorities.

Management strategies focused on carbon sequestration influence the climate in a variety of ways. Directly, carbon farming practices alter on-site $CO_2$ uptake and GHG emissions by the site's vegetation and soils. These direct changes determine the degree to which the local ecosystem sequesters carbon and functions as a GHG source or sink. In addition, management activities may entail emissions from machinery, materials production and transport, livestock, and other life-cycle processes (Cusack et al. 2021). These emissions may not occur on site but are nevertheless a meaningful component of a strategy's overall effect on the climate. Management actions in one location can also have indirect effects on GHG emissions or carbon sequestration elsewhere as a result of economic effects, material or labor availability, or management incentives (Murray et al. 2007). These indirect effects can be particularly hard to quantify, but merit consideration to assess whether carbon or GHG benefits are truly additional over a baseline counterfactual. Finally, the timescale over which carbon benefits are sustained depends on a system's long-term capacity to continue accruing carbon and the likelihood of reversal (Galik and Jackson, 2009), i.e., whether sequestered carbon will be re-emitted to the atmosphere in the event of disturbance or change in management practice.

Beyond their benefits for climate change mitigation, carbon management practices can lead to a suite of other outcomes for ecosystems and people. These potential co-benefits and tradeoffs include changes to water quality and supply, native vegetation and wildlife biodiversity, forage production and rangeland access for ranchers, soil health, and wildfire risk (e.g., Bullard and Smither-Kopperl, 2020; Dahlgren et al., 1997; Gravuer and Gunasekara, 2016; Russell and McBride, 2003; Salemi et al., 2012; Seavy et al., 2009; (Bullard and Smither-Kopperl; Tisdall and Oades, 1982). Some management actions may influence an ecosystem's resilience to climate change, and may have biophysical effects on local climate conditions due to shading, albedo changes, or altered rates of evapotranspiration (latent heat) (Pielke and Avissar, 1990). In some cases these benefits and tradeoffs are well understood, but in other cases they are challenging to predict due to insufficient data, high inherent variability, or complicated feedbacks and interactions. Finally, unintended effects such as pathogen introductions or fire can alter ecosystem structure, function, and potentially carbon stocks.

This chapter offers an overview of six management practices for the Alameda Watershed that managers might consider for carbon and GHG benefits. This list includes compost applications on the watershed's rangelands, riparian forest restoration, silvopasture, cattle exclusion to promote woody vegetation expansion, native grassland restoration, and conservation of the watershed as protected open space. Some of these practices are currently included in the management portfolio, whereas others would be new additions. Each of this chapter's sections highlights one of these practices, providing general background information, an estimate of potential carbon and GHG benefits based on models and/or published data, and an overview of co-benefits, tradeoffs, and other considerations regarding benefits, feasibility, and siting. To contextualize the carbon and GHG benefits of a given strategy, these rates are converted to approximate acreages that would be needed to offset 1% of San Francisco's 1990 GHG emissions of 8 MMT $CO_2e$/yr, or 80,000 MMT $CO_2e$/yr (City and County of San Francisco, 2021). (The San Francisco Climate Action Plan calls for nature-based solutions to offset up to 10% of San Francisco's 1990 emissions, requiring as much as ten times the acreages reported in this chapter; City and County of San Francisco, 2021.) Together, this information is intended to facilitate comparisons among the practices, offering managers the science background needed to make informed decisions that are compatible with SFPUC's management goals.

# RANGELAND COMPOST

## CARBON & CLIMATE BENEFITS

Sequesters 0.09–0.3 MT C/acre per year if compost is reapplied every 10 years.

Offsets GHG emissions by 0.3–1.1 MT $CO_2e$/acre per year if reapplied every 10 years.

With decadal reapplications, between 70,000 and 300,000 acres are needed to offset 1% of San Francisco's 1990 emissions.

Annual carbon and GHG benefits are lower with less frequent applications.

## CO-BENEFITS

AGRICULTURE   CULTURE   NATIVE BIODIVERSITY

RECREATION   SOIL QUALITY   WASTE MANAGEMENT

WATER QUALITY   WATER SUPPLY   WILDFIRE

## TRADEOFFS

AGRICULTURE   NATIVE BIODIVERSITY

WATER QUALITY   WILDFIRE

## ASSESSMENT

Numerous studies have demonstrated that one-time compost applications can sequester carbon in rangeland soils and provide life-cycle GHG benefits while increasing soil health and buffering against drought. Frequent reapplications may provide sustained benefits over multiple decades, but such long-term carbon sequestration rates are less well understood. Because limited research is available on other potential long-term effects such as vegetation community changes, pilot-scale projects on sites with low native vegetation cover offer a good choice for rangeland compost on the Alameda Watershed.

For millennia, humans have known that applying organic matter to soils can enhance soil health and increase fertility. Organic amendments increase soil organic matter (SOM) storage, decrease soil compaction, increase water holding capacity, reduce acidity, increase nutrient contents, enhance vegetation productivity, and improve other metrics of soil health such as microbial activity (del Mar Montiel-Rozas et al., 2016; Diacono and Montemurro, 2011; Hernando et al., 1989; Owen et al., 2015; Ryals et al., 2014, 2016). In recent years, compost applications have received substantial attention for climate change mitigation, as additions of composted organic material can increase soil carbon storage for years or decades with minimal associated GHG emissions (DeLonge et al., 2013; Mayer and Silver, 2022; Ryals and Silver, 2013). Organic matter amendments increase soil carbon storage in two fundamental ways. First, compost applications provide a direct source of carbon that slowly degrades and can be incorporated into SOM (Ryals et al., 2014). Over the longer term, increased rates of net primary production (NPP) due to compost applications have the potential to increase carbon storage for decades or more (Mayer and Silver, 2022; Owen et al., 2015).

With grasslands covering over 9% of California's land area, applying compost to rangelands has been included in the state's portfolio of strategies for climate change mitigation on natural and working Lands (CARB, 2022a; Silver et al., 2018). The City of San Francisco's Climate Action Plan (City and County of San Francisco, 2021), identifies compost applications as one of the carbon sequestration approaches to meet the city's target of net zero emissions by 2040. With over 12,000 acres of grassland, the Alameda Watershed has one of the largest footprints of city-owned and managed land where compost could be applied at scale.

For organic soil amendments to mitigate GHG emissions, the choice of material is important. While adding nutrient-rich material can increase productivity and enhance soil carbon storage, organic amendments may also stimulate methane ($CH_4$) and nitrous oxide ($N_2O$) emissions, offsetting or eliminating the benefits of carbon sequestration (Owen et al., 2015; Powlson et al., 2011). For instance, the common practice of spreading livestock manure over rangelands is associated with sizable increases in $N_2O$ emissions (Owen et al., 2015), a potent greenhouse gas with 265 times the global warming potential of $CO_2$ (100-year global warming potential; Myhre et al., 2013). In contrast, the use of composted material has been shown to sequester carbon and stimulate plant production with only minor increases in $CH_4$ or $N_2O$ emissions (Mayer and Silver, 2022; Ryals and Silver, 2013). Such composted material acts as a slow-release fertilizer without adding a large pulse of available nitrogen to stimulate $N_2O$ production (Mayer and Silver, 2022).

## CARBON AND GHG BENEFITS

Field studies and models have demonstrated the net carbon and GHG benefits of compost applications (e.g., Mayer and Silver, 2022; Ryals et al., 2015; Ryals and Silver, 2013). A field trial in California found that three years after amending soils with 0.5 inches of compost, soil carbon storage was 5.6–7.2 MT C/acre greater than in unamended controls, with no significant increases in $CH_4$ or $N_2O$ emissions (Ryals and Silver, 2013). 65–88% of this additional carbon was derived directly from the compost, with the remaining 12–35% due to increased carbon uptake (a fertilization effect). Modeling studies suggest that this fertilization benefit can be sustained over time, long after a single application (Mayer and Silver, 2022); using data from seven field sites in California, a model-based analysis found that carbon and GHG benefits persist for more than 85 years after a single 0.25 inch-deep compost application. Cumulative GHG benefits peak after ~18 years at 2.8 ± 0.04 MT $CO_2$e/acre, and then decrease over the following decades (Fig. 4.1) due to ongoing decomposition and reduced effects of fertilization over time (Mayer and Silver, 2022).

It has been hypothesized that re-applying compost at strategically timed intervals could maintain high rates of carbon sequestration over decades or longer (Mayer and Silver, 2022). To evaluate this potential, the CALAND model (Di Vittorio et al., 2021) was used to simulate the long-term carbon and GHG benefits of repeated compost applications on the Alameda Watershed. According to CALAND predictions, applying 0.5 inches of compost to the watershed's grasslands every 10 years over a period of 50 years could sequester an average of 0.2 MT C/acre per year over the coming decades (Fig. 4.1). This carbon benefit would decrease with less frequent applications, sequestering an estimated 0.05 MT C/acre per year if compost were reapplied every 30 years. Converting these carbon sequestration rates to units of $CO_2$e and accounting for GHG emissions due to compost production and application (DeLonge et al., 2013), these carbon sequestration rates translate to average GHG benefits of 0.7 (0.3–1.2) MT $CO_2$e/acre per year for 10-year applications or 0.2 (0.06–0.3) MT $CO_2$e/acre per year for 30-year applications. More frequent application rates may provide greater GHG benefits (Cameron et al., 2017), but this has not been tested rigorously with field studies or models.

Finally, it should be noted that carbon sequestration from compost applications is not unlimited; soil carbon may saturate at high carbon densities (Six et al., 2002), and carbon gains may be reversible if management practices are not maintained (Powlson et al., 2011).

## SITE CONSIDERATIONS

Within the Alameda Watershed, grassland sites with relatively low-grade slopes are likely the most feasible and appropriate sites for compost application projects (Fig. 4.2). Such sites are less likely than steeper sites to increase nutrient runoff into surface water, and generally offer safer machinery

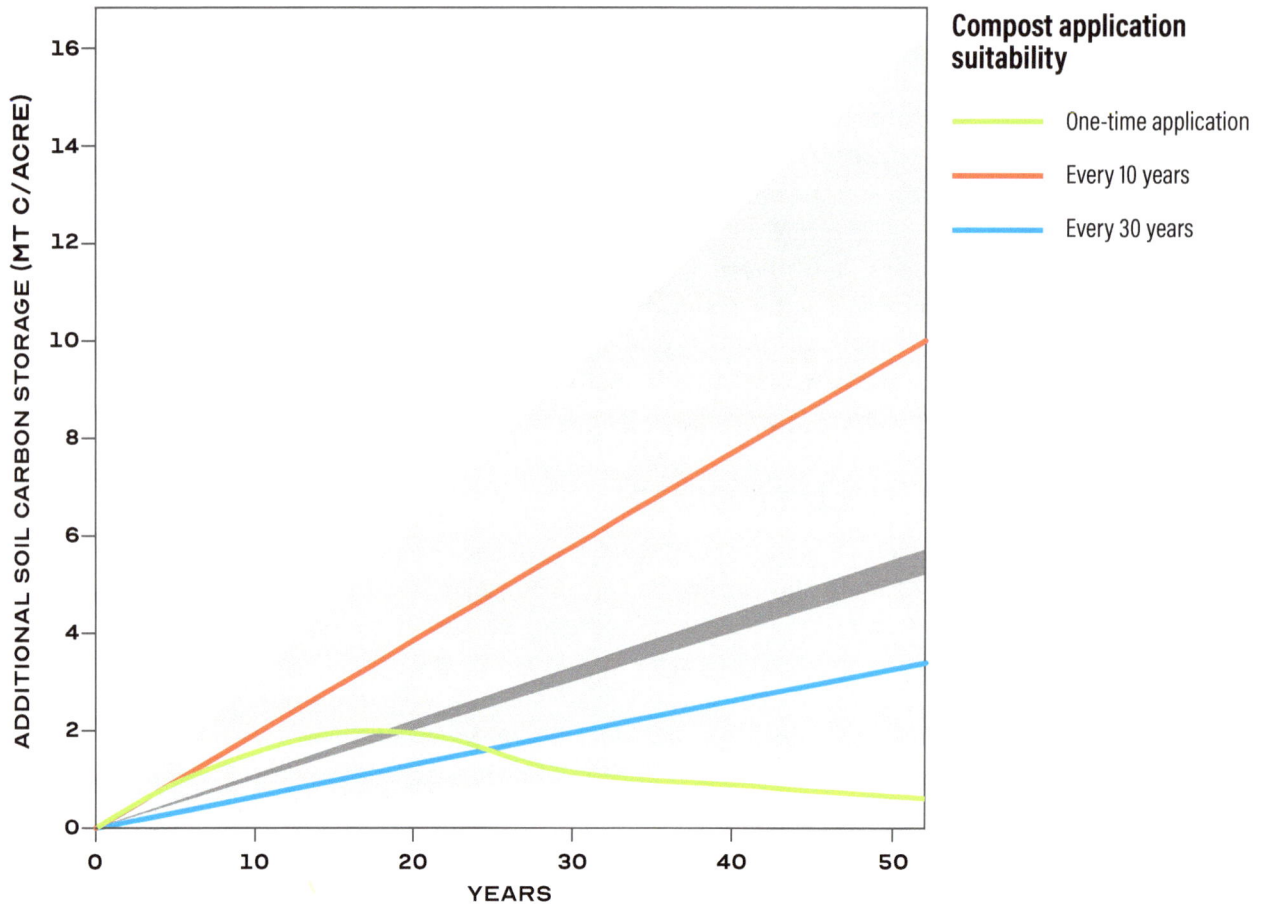

**Figure 4.1. Cumulative carbon benefits over time (MT C/acre) of rangeland compost applications under three application regimes.** The green line represents predicted carbon accumulation following a 1-time, 1/4 inch deep compost application, approximating values reported in Mayer and Silver (2022). The red and blue lines represent carbon accumulation following repeated 1/2 inch compost applications at 10-year (red) or 30-year (blue) intervals, as predicted by the CALAND model. Gray shading represents the uncertainty band reported by CALAND, which is based on uncertainties in initial carbon densities and historical carbon fluxes.

access (Gravuer and Gunasekara, 2016)—though low-grade sites far from roads may present access challenges. The NRCS conservation practice standard for soil carbon amendments recommends that compost only be applied to ranglands with slopes of 8% or less (NRCS, 2020), and the San Mateo and Marin County Resource Conservation Districts recommend a maximum slope of 20% for compost applications. An ongoing Healthy Soils Demonstration Project through the Alameda County Resource Conservation District is evaluating the use of compost on sites in the Altamont Pass with up to 30% slopes (R. Ryals, pers. comm.). If successful, this would provide an example indicating that higher-grade slopes on the Alameda Watershed may be appropriate where accessible by machinery.

In addition to topography, existing vegetation is an important consideration for rangeland compost projects. Organic soil amendments may be particularly beneficial for forage production and carbon sequestration in heavily disturbed grasslands with low vegetation cover or productivity (Ohsowski et al., 2012). In contrast, sites with high native vegetation cover or locally rare species or communities may not be appropriate for compost applications, due to unknown effects of long-term nutrient additions on native plant diversity (Gravuer et al., 2019). The effect of soil nutrients on individual species has been shown to vary from site to site, even at relatively small spatial scales within Cal-

N

| 4 miles |
| 4 km |

**Oak savanna and non-serpentine grassland on slopes less than 30%**

Slope <8%: 1,875 acres

Slope 8-20%: 4,448 acres

Slope 20-30%: 3,994 acres

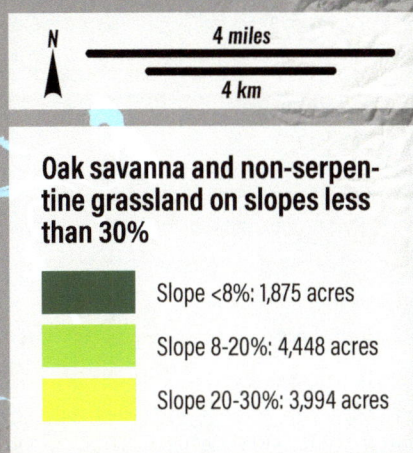

**Figure 4.2. A starting point for identifying the potential opportunity space for rangeland compost application in the Alameda Watershed.** Sites represented in the mapping include oak savanna and non-serpentine grasslands on slopes less than 30%, as well as the former Sunol Valley Golf Course. In addition to vegetation type and slope, a number of other factors should be considered to identify sites where rangeland compost application might be an appropriate management strategy, such as proximity to drinking water reservoirs, native plant cover, presence of rare and endangered plant and animal habitat, and site accessibility. As noted above, there are major concerns associated with compost application, and the mapping does not assess the desirability of applying this strategy within potential opportunity areas.

ifornia's Diablo range (Spiegal et al., 2014), making it challenging to predict how nutrient additions will influence native and nonnative vegetation. One study from the northern Diablo range found that low-nitrogen soils may provide a refuge for native species within invaded California grasslands (Gea-Izquierdo et al., 2007), suggesting that there is reason for concern that compost additions could negatively impact native vegetation communities on the watershed. The NRCS standards suggest a cautious approach, stating that compost application projects should avoid sites dominated by native, special status, or locally rare vegetation, such as serpentine grasslands, coastal prairie, chaparral, or coastal sage scrub ecosystems (NRCS, 2020). Similarly, compost is not recommended on sites containing special status native plants and/or animals that require low-stature rangeland habitat, such as the California tiger salamander (*Ambystoma californiense*) and burrowing owl (*Athene cunicularia*) (Gravuer and Gunasekara, 2016).

The former Sunol Valley Golf Course is an example site with a high degree of historical disturbance, generally flat topography, and low native vegetation diversity. These characteristics suggest that it could be a good choice for pilot compost applications. If historical turf management included high fertilizer additions, however, this could increase the potential for nutrient runoff or $N_2O$ emissions. If the former golf course is used for pilot-scale compost application trials, careful monitoring of runoff chemistry and gaseous emissions may be needed.

## OTHER CONSIDERATIONS

The GHG benefits of compost applications depend on the characteristics of organic material and its alternative fate (Powlson et al., 2011). Compost applications can provide additional life-cycle GHG benefits if they divert waste from traditional waste management systems with high GHG emissions (DeLonge et al., 2013). Assessing these life-cycle benefits requires a careful accounting of counterfactual conditions. A life-cycle analysis of a California rangeland compost project, for example, found that avoided GHG emissions from landfill and manure slurry systems were greater than on-site $CO_2$ sequestration (DeLonge et al., 2013). On the other hand, if the material would otherwise be composted and used elsewhere, on-site carbon sequestration and life-cycle GHG benefits may not be truly *additional*. In the case of the Alameda Watershed, potential composting projects should consider how the material is processed, how it would otherwise be used, and whether applying compost to the watershed would increase the total footprint of rangeland compost projects across the state.

Much of the research on rangeland compost applications has focused on composted livestock manure and green waste. Alternatively, class A or B biosolids from wastewater treatment plants offer another source of organic material for grassland composting projects. Studies in California and elsewhere have documented soil carbon sequestration due to biosolids application (Brown et al., 2011; Pan et al., 2017; Villa and Ryals, 2021; Wijesekara et al., 2017), and the SFPUC Wastewater Enterprise supplies processed biosolids to California ranchers for use as an organic fertilizer (https://sfpuc.org/programs/biosolids). For the purpose of climate change mitigation, however, biosolids application may not represent a net win; the high nitrogen content typical of digested or composted biosolids (C:N ratios around 4–8; Pan et al., 2017; Tian et al., 2015; Villa and Ryals, 2021) has the potential to stimulate high grassland $N_2O$ emissions and lead to vegetation community changes. A global meta-analysis of organic soil amendments classified biosolids as "high-risk" for $N_2O$ emissions (Charles et al., 2017), and the California Department of Food and Agriculture (CDFA) Healthy Soils Incentives Program recommends that compost applied to rangelands have a C:N ratio of 11 or above in order to mitigate changes to vegetation communities (Gravuer and Gunasekara, 2016). For the purpose of GHG mitigation, it is thus not recommended to apply biosolids alone, i.e., without additional amendments to increase the C:N ratio.

The potential for compost additions to enhance rangeland carbon storage may have interacting effects with wildfire, warming, drought, and other climate-related factors. In some ways, compost applications may enhance sites' resilience to the effects of climate change. In addition to buffering vegetation productivity against water stress in dry years (Ryals and Silver, 2013), organic matter amendments have been demonstrated to restore soil fertility following fire (Cellier et al., 2014), though recently burned sites are generally ineligible for compost applications under California's Healthy Soils program (Gravuer and Gunasekara, 2016). At the same time, however, compost can reduce grassland plant diversity and favor nonnative species (Bullard and Smither-Kopperl, 2020; Gravuer et al., 2019; Seabloom et al., 2021; Suding et al., 2005), and potential carbon gains from compost may decrease with warming due to increased microbial decomposition rates (Mayer and Silver, 2022). Another consideration with compost additions is the potential for plant pathogen introductions if compost made from green waste is not produced and handled properly.

| CATEGORY | COSTS AND BENEFITS |
|---|---|

**AGRICULTURE**

**Co-benefits**

Compost additions can increase production of forage for livestock, particularly in heavily disturbed grasslands (Bullard and Smither-Kopperl, 2020; Ohsowski et al., 2012). In years with unusually low precipitation, organic matter additions can buffer forage production rates by increasing soil water retention (Ankenbauer and Loheide, 2017; Lal, 2020; Ryals and Silver, 2013).

**Tradeoffs**

Increases in forage production associated with compost applications may require rangeland managers to alter their standard practices to maintain optimal residual dry matter loads. Each cow grazed on the Alameda Watershed emits an estimated 0.7–1.5 MT $CO_2e$ per year as methane due to enteric fermentation (IPCC, 2006). If improved forage conditions increased cattle stocking rates according to guidance in the draft Alameda Creek Watershed Rangeland Management Plan for unfavorable vs. normal forage conditions (from 9.67 to 4.05 acres per cow-calf pair; SFPUC, 2017), increased methane emissions would reduce the GHG benefit of compost by ~14%.

**NATIVE BIODIVERSITY**

**Tradeoffs**

Nutrient additions often reduce grassland plant diversity and can benefit invasive species by reducing the diversity of environmental controls (Davis et al., 2000; Harpole et al., 2016; Seabloom et al., 2021; Suding et al., 2005). Studies evaluating the effects of compost on rangeland vegetation communities have reported mixed results, in some cases finding that fertilization favors nonnative annual grasses (Bullard and Smither-Kopperl, 2020), and in other cases finding minimal effects to vegetation communities (Ryals et al., 2016) or variable results across studies (Gravuer et al., 2019). Given this uncertainty, the CDFA Healthy Soils Incentives Program and NRCS conservation practice standard recommend that compost amendment projects avoid native grasslands, sites that have high concentrations of rare species, and/or sites that may otherwise be particularly sensitive to nutrient additions (Gravuer and Gunasekara, 2016; NRCS, 2020).

| CATEGORY | COSTS AND BENEFITS |
|---|---|

**SOIL QUALITY**

**Co-benefits**

Organic matter amendments enhance soil fertility by decreasing soil compaction, raising soil pH, increasing nutrient availability, and improving aggregation and water retention (Diacono and Montemurro, 2011; Hargreaves et al., 2008; Hernando et al., 1989; Tisdall and Oades, 1982).

**WASTE MANAGEMENT**

**Co-benefits**

The production and beneficial reuse of composted organic materials can divert material from traditional waste management streams (Brown et al., 2008; DeLonge et al., 2013).

**WATER QUALITY**

**Co-benefits**

Organic matter inputs can increase soil water holding capacity and improve infiltration, reducing runoff and improving surface water quality (Brown and Cotton, 2011).

**Tradeoffs**

In addition to their potential water quality benefits, compost applications also have the potential to negatively affect water quality through increased nutrient runoff. Moderate compost application rates and low-slope sites are recommended to limit negative water quality effects (Gravuer and Gunasekara, 2016).

**WILDFIRE**

**Tradeoffs**

In other western rangelands, herbaceous fuel load and fuel moisture have been seen to influence fire intensity (Davies et al., 2015). If increased grassland production associated with compost additions is not effectively managed through grazing, increased fuel loads may increase the risk of high-intensity, fast-spreading wildfire. No studies, however, have evaluated the potential for compost amendments to increase wildland fuel loads or wildfire risk.

# RIPARIAN FOREST RESTORATION

## CARBON & CLIMATE BENEFITS

Revegetating riparian forest areas sequesters 30–50 MT C/acre over 50 years.

Averaged over 50 years, this offsets GHG emissions by 3–4 MT $CO_2$e/acre per year.

Revegetating 20,000–40,000 acres of riparian forest will offset 1% of San Francisco's 1990 emissions for a period of 50 years.

## CO-BENEFITS

AGRICULTURE    CULTURE    NATIVE BIODIVERSITY

RECREATION    SOIL QUALITY    WASTE MANAGEMENT

WATER QUALITY    WATER SUPPLY    WILDFIRE

## TRADEOFFS

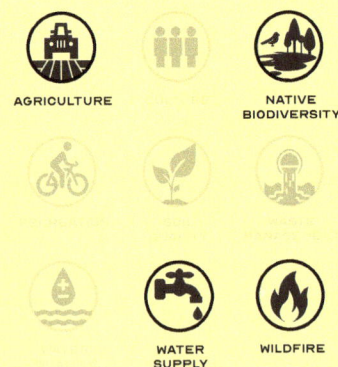

AGRICULTURE    NATIVE BIODIVERSITY

WATER SUPPLY    WILDFIRE

## ASSESSMENT

Planting trees and excluding cattle in riparian areas offers high rates of carbon sequestration as well as other co-benefits for biodiversity and water resources.

Several types of forested riparian habitats occur within the Alameda Watershed. Coast live oak (*Quercus agrifolia*)-dominated riparian forests are the most common (420 acres), found in canyons and floodplains along both intermittent and perennial streams. Sycamore alluvial woodland (dominated by *Platanus racemosa*; 342 acres) occurs along low gradient intermittent streams characterized by flood disturbance during the wet season and lack of water during the dry season. White alder riparian forest (dominated by *Alnus rhombifolia*; 139 acres mapped) occurs along perennial streams such as Alameda Creek and Arroyo Hondo, while willow riparian forest/scrub (dominated by *Salix* spp.) occurs near the active channel of perennial and intermittent streams (SFPUC, 2015). The mapped extent of forested riparian habitats is likely an underestimate, as detection in aerial photographs can be difficult. Non-forested riparian habitats are also common, particularly within smaller drainages in the upper watershed.

The extent of riparian forest within the watershed has likely diminished over time as a result of changes in land use and hydrology. In southern Sunol Valley, for instance, broad corridors of sycamore alluvial woodland, in some places more than 400 m wide, existed historically on the floodplains surrounding Alameda Creek. Further upstream, Alameda Creek supported a similarly broad corridor of riparian woodland characterized by willows (*Salix* spp.), oaks (*Quercus* spp.), sycamores (*Platanus racemosa*), and alders (*Alnus* spp.; Stanford et al. 2013). Today, large portions of historical riparian corridors along larger channels in valley floors have been converted to other land uses such as aggregate mining and nurseries (SFPUC, 2015). Dam construction and other hydromodifications have likewise impacted riparian habitats in some parts of the watershed.

Restoration of riparian forests has the potential to increase levels of ecosystem carbon storage substantially. Existing riparian forests have the highest density of biomass carbon of any ecosystem type in the Alameda Watershed (see page 29), and studies have shown that riparian forest restoration can help sequester carbon in vegetation and soils while supporting biodiversity (Dybala et al., 2019; Golet et al., 2008; Matzek et al., 2020; Seavy et al., 2009), reducing erosion and increasing infiltration (Brauman et al., 2007; Gyssels et al., 2005), and filtering nutrients from runoff (Dosskey et al., 2010; Mayer et al., 2005). A global synthesis and meta-analysis found that riparian restoration increased soil carbon stocks by more than 200%

on average, while total biomass carbon storage in the restored forests ranged from 28-64 MT C/acre (Dybala et al., 2019). Soil carbon in restored riparian habitats tends to be highest in upper bank terrace positions (further from the channel), while vegetation carbon tends to be highest in depositional floodplain positions (Lewis et al., 2015; Matzek et al., 2020). Active planting can more than double initial growth rates relative to natural riparian forest regeneration and substantially decrease time needed to reach maturity (Dybala et al., 2019).

## CARBON AND GHG BENEFITS

The Carbon in Riparian Ecosystems Estimator for California (CREEC) tool (Matzek et al., 2018) was used to estimate levels of carbon sequestration for three different restoration scenarios (natural riparian regeneration, planted willow riparian, and planted oak or sycamore riparian) on previously grazed lands in the Alameda Watershed, with a time horizon of up to 100 years (Fig. 4.3). Among the riparian restoration and management activities practiced on the Alameda Watershed, these scenarios represent cattle exclusion, tree planting (in the case of the two planted scenarios), and protection of new plantings from large herbivores. The CREEC calculator assumes that sites transition over time from non-forested to forested due to restoration activities.

For the natural regeneration scenario, modeled carbon accumulation equaled 47 MT C/acre over the first 50 years, with an average 50-year carbon sequestration rate of 0.93 MT C/acre-yr (3.4 MT $CO_2$e/acre-yr). Accumulation rates differed for scenarios with planted riparian communities; over the first 50 years following restoration, mean annual carbon sequestration equaled 0.61 MT C/acre-yr for willow riparian and 1.0 MT C/acre-yr for oak/sycamore riparian vegetation. Results from the CREEC tool show that carbon accumulates rapidly within the first 20–40 years following restoration, with peak carbon sequestration rates occurring between 10 and 20 years (Fig. 4.3). After the first few decades, carbon continues to accumulate more gradually in the absence of major disturbances. For instance, estimates for cumulative C uptake with natural regeneration ranged from 14.44 to 15.21 MT C/acre over the first 10 years, 30.42 to 49.54 over 50 years, and 34.03 to 53.79 over 100 years. These findings are consistent with Dybala et al. (2019), who found that within three decades, carbon storage in both planted and naturally regenerating riparian forests approached that of remnant forest stands.

This analysis assumes that cattle exclusion and planting are sufficient to establish riparian forest vegetation. In practice, woody riparian vegetation may be challenging to restore, given other influences such as modified hydrology below dams and near quarries, drought and warming associated with climate change, herbivory, and challenges with restoring native plant communities in degraded and invaded sites. In such cases, lower restoration success would yield reduced carbon and GHG benefits. Additionally, even when restoration is successful, the potential GHG mitigation benefit of riparian restoration can be partially offset by short-term carbon losses and GHG emissions under certain conditions. Soil carbon stocks may initially decrease following restoration as a result of soil disturbance and increased decomposition rates (Mackay et al., 2016). If riparian restoration increases floodplain inundation frequency through modifications to the channel structure, more frequent flooding can increase $CH_4$ and $N_2O$ emissions (Dybala et al., 2019; Jacinthe, 2015; Jacinthe et al., 2015; Welsh et al., 2021). Furthermore, a portion of the carbon stored by riparian restoration may be due to sediment deposition during flood events, increased runoff filtering,

**RIPARIAN FOREST**

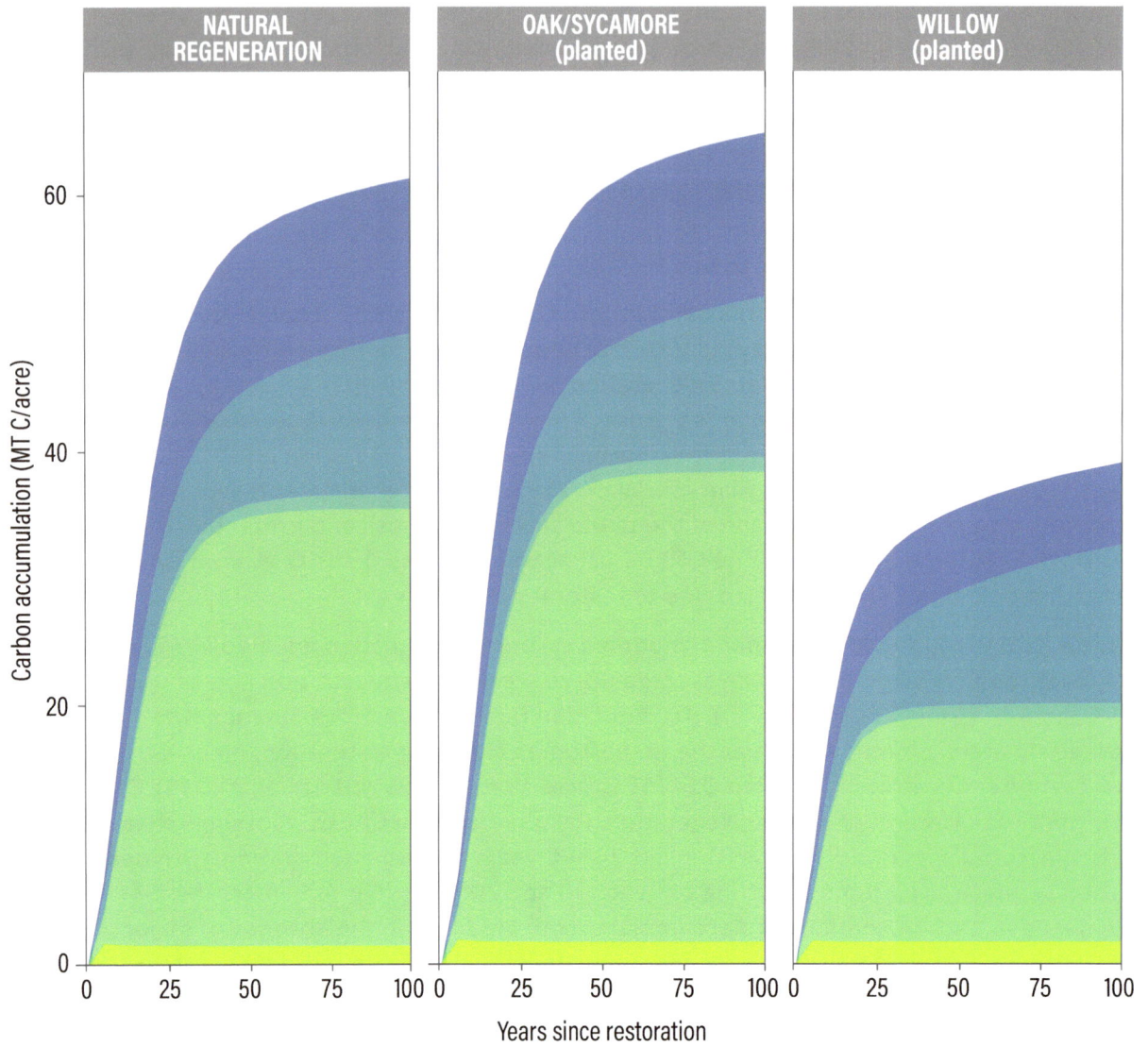

Carbon accumulation (MT C/acre)

NATURAL REGENERATION

OAK/SYCAMORE (planted)

WILLOW (planted)

Years since restoration

**Carbon Pool**

- Downed dead wood
- Forest floor
- Soil
- Standing trees
- Understory

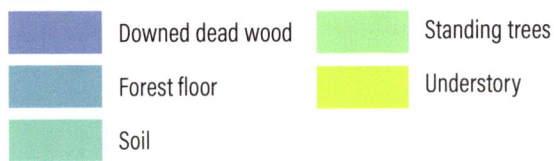

**Figure 4.3. Carbon accumulation over time in restored riparian forest communities.** Carbon accumulation curves are based on predictions from the CREEC estimator (Matzek et al., 2018) for natural regeneration or planted vegetation communities typically used in riparian restoration projects on the Alameda Watershed.

or reduced erosion. Such carbon represents net sequestration only if it would otherwise have been more rapidly respired (Berhe et al., 2007; Dybala et al., 2019).

## SITE CONSIDERATIONS

While many factors influence the suitability of a particular site for riparian forest restoration (e.g., slope and aspect, streamflow patterns, soil characteristics), at a first pass the potentially suitable sites for riparian restoration in the Alameda Watershed are likely to be currently unforested areas adjacent to higher order streams (Fig. 4.4). The potential width of the riparian corridor generally increases with stream order: lower order streams may support a corridor width of just 30 m or less, while higher order streams may support a corridor more than 100 m wide (Central Coast Wetlands Group, 2017). Assuming that restoration of riparian forests is most feasible in unforested areas within 30-100 m of streams with Strahler order greater than 2, the potentially suitable area for riparian restoration in the Alameda Watershed is estimated at between 2,550 and 3,790 acres (depending on the land cover mapping source used to identify unforested areas). Selection of appropriate target riparian habitat types for each restoration site should be based on an understanding of historical ecology and contemporary site conditions.

Alameda Creek, courtesy of SFPUC.

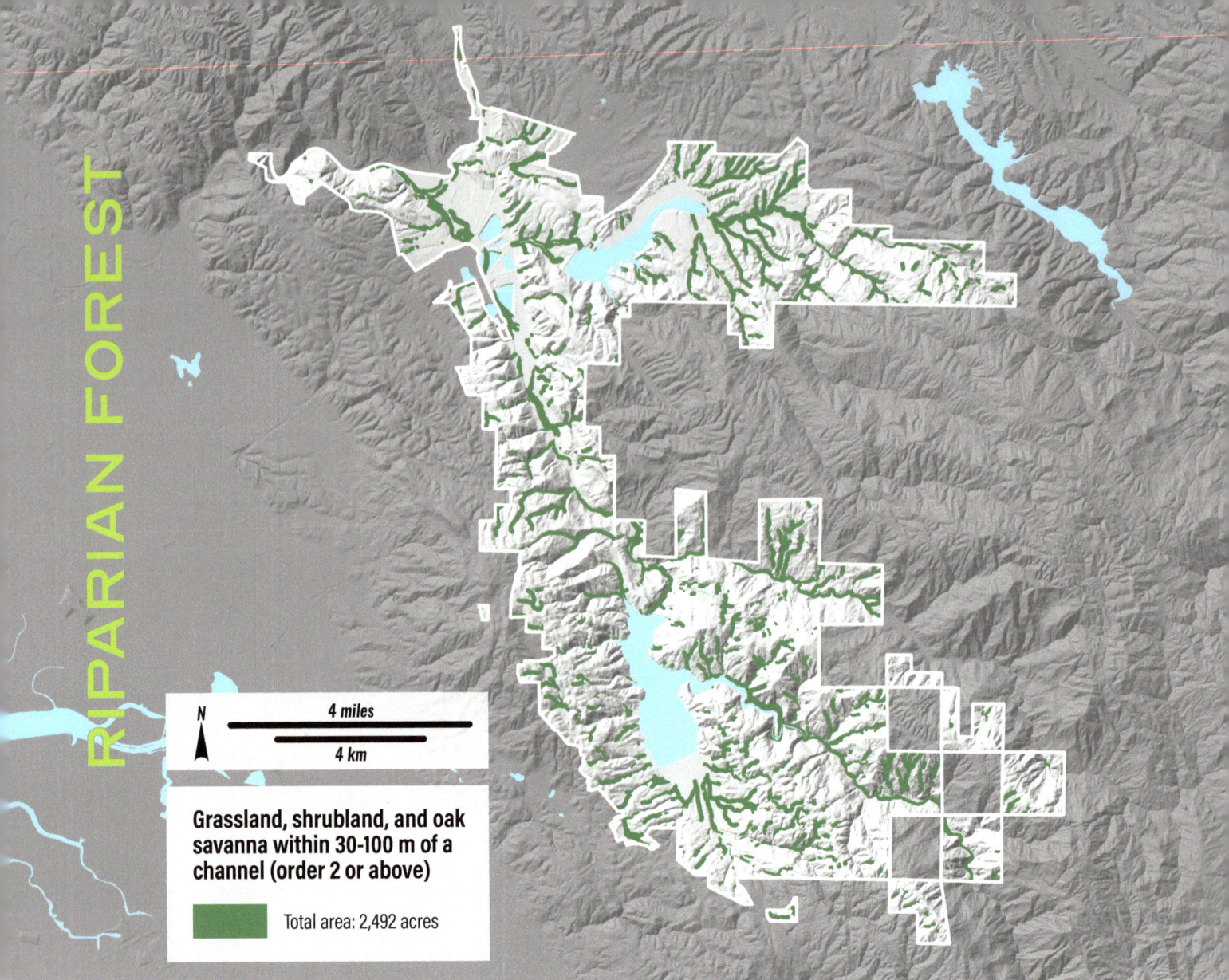

RIPARIAN FOREST

**N** | 4 miles
| 4 km

**Grassland, shrubland, and oak savanna within 30-100 m of a channel (order 2 or above)**

Total area: 2,492 acres

**Figure 4.4. A starting point for identifying potential opportunity space for riparian restoration projects in the Alameda Watershed.** Sites represented in the mapping include areas not classified as woodland by Jones and Stokes (2003) mapping that fall within 30–100 m of channels for stream orders 2 and above. Potential riparian buffer width depends on stream order, as described in the technical appendix. In addition to existing vegetation and potential riparian buffer width, other factors to consider in identifying sites where riparian restoration is an appropriate management strategy may include accessibility, slope and aspect, streamflow patterns, soil characteristics, and surrounding land use. Accessibility may determine the suitability of sites for riparian restoration: many riparian areas on the Alameda Watershed are steep and difficult to access, and riparian restoration requires materials such as fencing, cages, tubes, and protective sleeves, ongoing maintenance of plantings, and in some cases irrigation or water transport.

| CATEGORY | COSTS AND BENEFITS |
|---|---|

**AGRICULTURE**

**Co-benefits**

In rangeland settings, riparian forests provide local cooling and shading for livestock.

**Tradeoffs**

Excluding livestock to restore riparian vegetation can limit access to water sources and increase the need for water infrastructure elsewhere on grazed lands.

**NATIVE BIODIVERSITY**

**Co-benefits**

Riparian forests are used as habitat by a wide range of species, including birds, bats, other vertebrates, and invertebrates (Goals Project, 1999; Golet et al., 2008; Hilty and Merenlender, 2004; Lennox et al., 2011; Opperman and Merenlender, 2004). Riparian forests shade adjacent aquatic habitats and provide important resources for aquatic food webs. As linear corridors, riparian forests provide critical connectivity and facilitate wildlife movement through the landscape (Hilty and Merenlender, 2004). With their cooler and wetter microclimate relative to the surrounding landscape, riparian forests provide important refugia for wildlife, a function that will become increasingly important with accelerating climate change (Seavy et al., 2009).

**Tradeoffs**

Active planting of woody riparian species during restoration activities carries a risk of plant pathogen introduction if container plantings are used. Additionally, if cattle are no longer allowed access to natural water sources, creating alternative water infrastructure can impact ecosystems elsewhere, either through soil disturbance and vegetation removal, or in some cases via water diversions from natural seeps and springs.

| CATEGORY | COSTS AND BENEFITS |
|---|---|

**WATER QUALITY**

**Co-benefits**

Riparian corridors help to maintain water quality by buffering streams from nutrient, sediment, and pollutant inputs (Anbumozhi et al., 2005; Sweeney and Newbold, 2014). Riparian forests can lower rates of gross primary productivity and ecosystem respiration in stream habitats, helping to decrease eutrophication (Burrell et al., 2014; Zefferman, 2014).

**WILDFIRE**

**Tradeoffs**

Riparian vegetation typically burns less frequently than upland vegetation in the central California foothills region (Bendix and Commons, 2017), and can act as a natural fire break due to higher moisture contents than surrounding vegetation (Kobziar and McBride, 2006). In some cases, however, riparian forests have the potential to act as conduits for wildfire across the landscape during severe fires (Pettit and Naiman, 2007). Restoration of woody riparian vegetation could thus lead to increased risk of wildfire along riparian corridors during dry periods.

**SOIL QUALITY**

**Co-benefits**

Riparian forests can enhance soil quality through stabilization (Gyssels et al., 2005), litterfall, and nutrient capture (Matzek et al., 2016).

**WATER SUPPLY**

**Tradeoffs**

Restoring riparian forest vegetation has the potential to reduce water yield, as demonstrated in numerous studies evaluating vegetation removal, vegetation restoration, and paired vegetated and unvegetated riparian sites (reviewed in Salemi et al., 2012).

# SILVOPASTURE

## CARBON & CLIMATE BENEFITS

Increasing oak density in grassland and savanna sequesters 3 MT C per tree over 60 years.

Per-acre benefits of silvopasture depend on stand density. Averaged over 60 years, a density of 6 trees per acre would offset GHG emissions by 1.3 MT $CO_2$e/acre per year.

An estimated 380,000 additional trees would offset 1% of San Francisco's 1990 emissions for a period of 60 years.

## ASSESSMENT

Planting trees in rangelands—a form of silvopasture—can sequester carbon and enhance soil fertility in open grasslands, benefiting livestock and vegetation communities. In the Alameda Watershed, silvopasture approaches that reflect historical native ecosystem patterns include the expansion of savanna or open-canopy oak woodland. The long-term carbon benefits of silvopasture depend on the survival of planted trees, which face multiple environmental stressors (drought, wildfire, etc.) related to climate change.

## CO-BENEFITS

AGRICULTURE    CULTURE    NATIVE BIODIVERSITY

RECREATION    SOIL QUALITY    WASTE MANAGEMENT

WATER QUALITY    WATER SUPPLY    WILDFIRE

## TRADEOFFS

AGRICULTURE    NATIVE BIODIVERSITY

WATER SUPPLY

Silvopasture is an agroforestry management strategy that includes the addition of trees in grazed rangeland to increase an area's overall productivity. Silvopasture can provide multiple benefits for ecosystems and people, including shade and shelter for livestock, water quality improvements, wildlife habitat, enhanced forage production, carbon sequestration, and erosion mitigation (USDA NRCS, 2016), and has also been identified as one of the most well-supported management practices for rangeland soil health (Carey et al., 2020).

The ideal configuration and composition of silvopasture systems depends on local ecological conditions and management goals. The general recommendation from the Natural Resources Conservation Service (NRCS conservation practice standard 381) is to select species that are adapted to the site's conditions and compatible with its management goals, and to maintain a tree stocking density of at least 10% (USDA NRCS, 2016). In many systems, silvopasture is managed not only for forage but also for forest or agricultural products, favoring timber species such as pine or agricultural species such as fruit or nut trees. In the Alameda Watershed, expansion of savanna or open-canopy oak woodland represents an approach to silvopasture that could offer carbon and GHG benefits while maintaining natural communities native to the watershed. Historically, oak savanna ecosystems occupied many alluvial valleys in the Alameda Watershed and were typically dominated by relatively low densities of valley oak and blue oak (Stanford et al., 2013). Tree densities in some historical savanna settings, such as La Costa Valley upstream of San Antonio Reservoir, have declined over time as a result of tree clearing for agriculture and other uses (Fig. 4.6; Stanford et al., 2013). In such settings, converting open grasslands to low-density hardwood rangeland could provide the various benefits of silvopasture systems in a way that is compatible with the region's historical ecology.

## CARBON AND GHG BENEFITS

Carbon benefits associated with tree plantings accrue cumulatively over time, and the rate of annual carbon uptake changes over time with tree growth. To estimate potential carbon sequestration associated with silvopasture, the i-Tree calculator was used to quantify biomass carbon accumulation in individual valley oaks planted in Sunol. Trees were assumed to be spaced such that there were no effects of shading or other competition from surrounding trees, and were assumed to survive to maturity. Carbon accumulation in downed dead wood was calculated using the constant factor of 0.062 x standing tree carbon reported in Matzek et al. (2018) for oak-dominated sites. Results indicate that each newly-established valley oak in the Alameda Watershed would accumulate an estimated 0.1 MT C over the first 10 years, 3 MT C over 60 years, and 5 MT C cumulatively over a century. Averaged over the 60 years following seedling establishment—the period of most rapid carbon accumulation (Fig. 4.5)—model-predicted annual C sequestration equaled 0.06 MT C/tree, providing an average GHG benefit of 0.2 MT $CO_2$e/tree-yr.

The carbon sequestration rate reported above includes standing aboveground biomass, roots, and downed dead wood, but does not account for carbon accumulation in understory vegetation, litter, or other accumulations around the base of the tree. These additional carbon pools would likely enhance the carbon benefit associated with silvopasture, but are challenging to quantify with existing

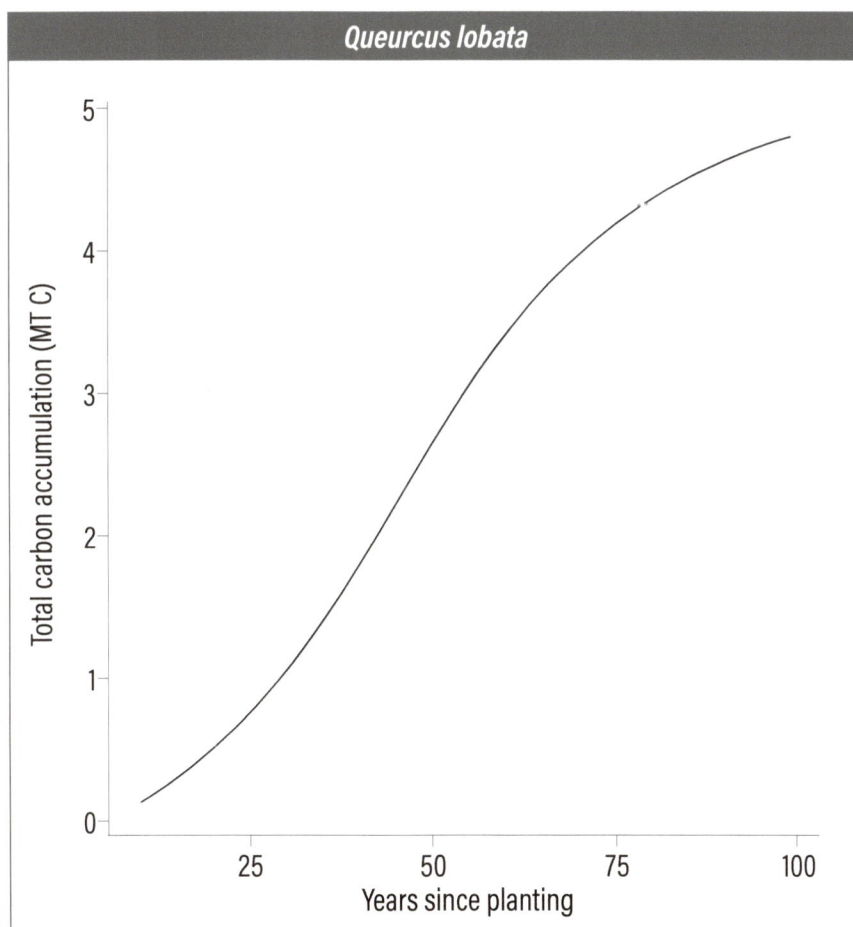

**Figure 4.5. Biomass carbon accumulation in individual *Quercus lobata* trees.** Carbon accumulation over time was based on the i-Tree Planting tool for *Q. lobata* in Sunol, CA, assuming no shading from nearby vegetation or other major disturbance events.

data or models. Additionally, silvopasture may also increase soil carbon storage, as suggested by numerous studies reporting higher soil organic carbon under oaks than in adjacent open grasslands (Camping, 2002; Dahlgren et al., 1997; Herman et al., 2003). Based on a synthesis of soil carbon data from open grassland and wooded sites (Fig. 3.3), establishing trees in the watershed's grasslands may increase soil carbon storage by an estimated 20 MT C/acre over the long term (80+ years, based on stand ages from soil carbon studies).

Scaling potential per-tree carbon accumulation to per-acre sequestration rates depends on planting density and survivorship of newly established trees. Based on i-Tree results, at a density of 1 tree per acre—comparable to 10% canopy cover characteristic of low-density savanna for fully established trees (Ch.6, Grossinger et al., 2008)—silvopasture could provide an average annual GHG benefit of 0.2 MT $CO_2$e/acre-yr over the 60 years following tree establishment, whereas a higher density of 6 trees per acre (sparse-canopy woodland with 60% cover) could provide an average annual GHG benefit of 1.3 MT $CO_2$e/acre-yr. These values pertain to final stand densities of living trees, which differ from initial planting densities due to mortality from water stress, herbivory, disease, fire, or other environmental stressors. Accordingly, actual carbon sequestration in silvopasture plantings should be tracked over time according to surviving trees, not extrapolated from initial plantings. Additionally, carbon accumulation rates decline after the first ~60 years, so maintaining estimated annual GHG benefits beyond this timeframe would require additional tree planting over time.

**Trees along Alameda Creek through lower Sunol Valley.** This early 20th century photograph shows scattered oaks along barely visible Alameda Creek. (SVWC; courtesy San Francisco History Center, San Francisco Public Library.)

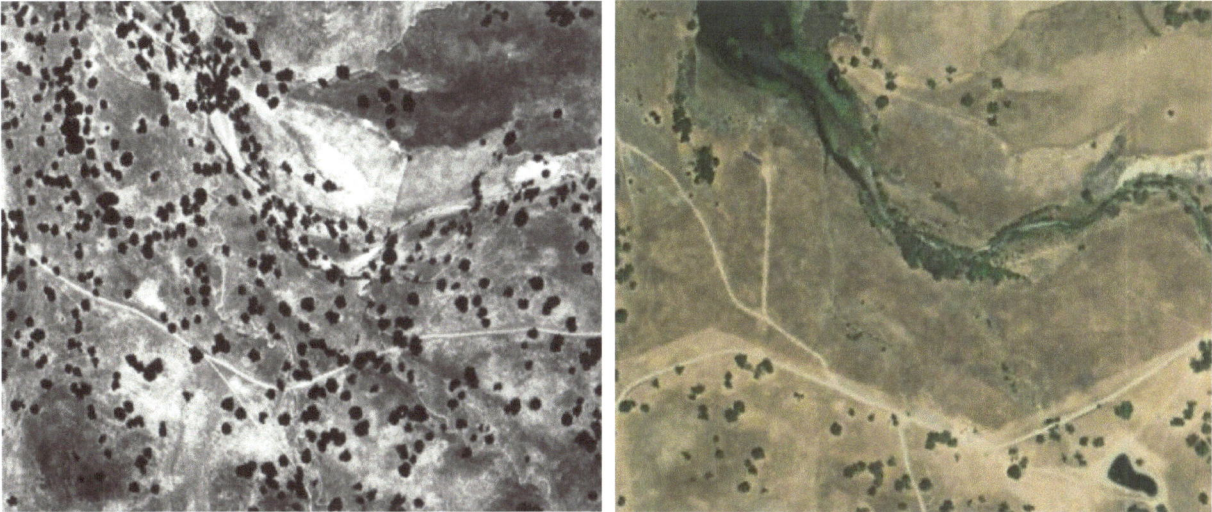

Figure 4.6. Historical change in oak density in La Costa Valley (just upstream of San Antonio Reservoir) between 1939-40 and 2021 (USDA 1939-40, 2021 imagery from ESRI).

Grasslands and oaks, Alameda Watershed, courtesy of SFPUC.

Over 100 years, silvopasture systems would accumulate an estimated 5 MT C/acre at 10% canopy cover, or 29 MT C/acre at 60% canopy cover. In comparison, existing oak savanna sites on the Alameda Watershed store an average of 18 MT C/acre in standing trees and downed dead material, or 21 MT C/acre in all biomass pools (Table 3.2). Achieving a comparable carbon gain through silvopasture would require establishing oaks in open grassland with a final stand density of ~37% canopy cover, or 3.8 trees/acre. For comparison, Fig. 4.7 shows a selection of existing savanna sites within the watershed with biomass carbon densities ranging from 13.5–21.8 MT C/acre. Converting regions of open grassland to similarly structured savanna would build carbon stocks over time by an estimated 11–19 MT C/acre while maintaining grazing access and providing shade for livestock.

## BIOMASS CARBON

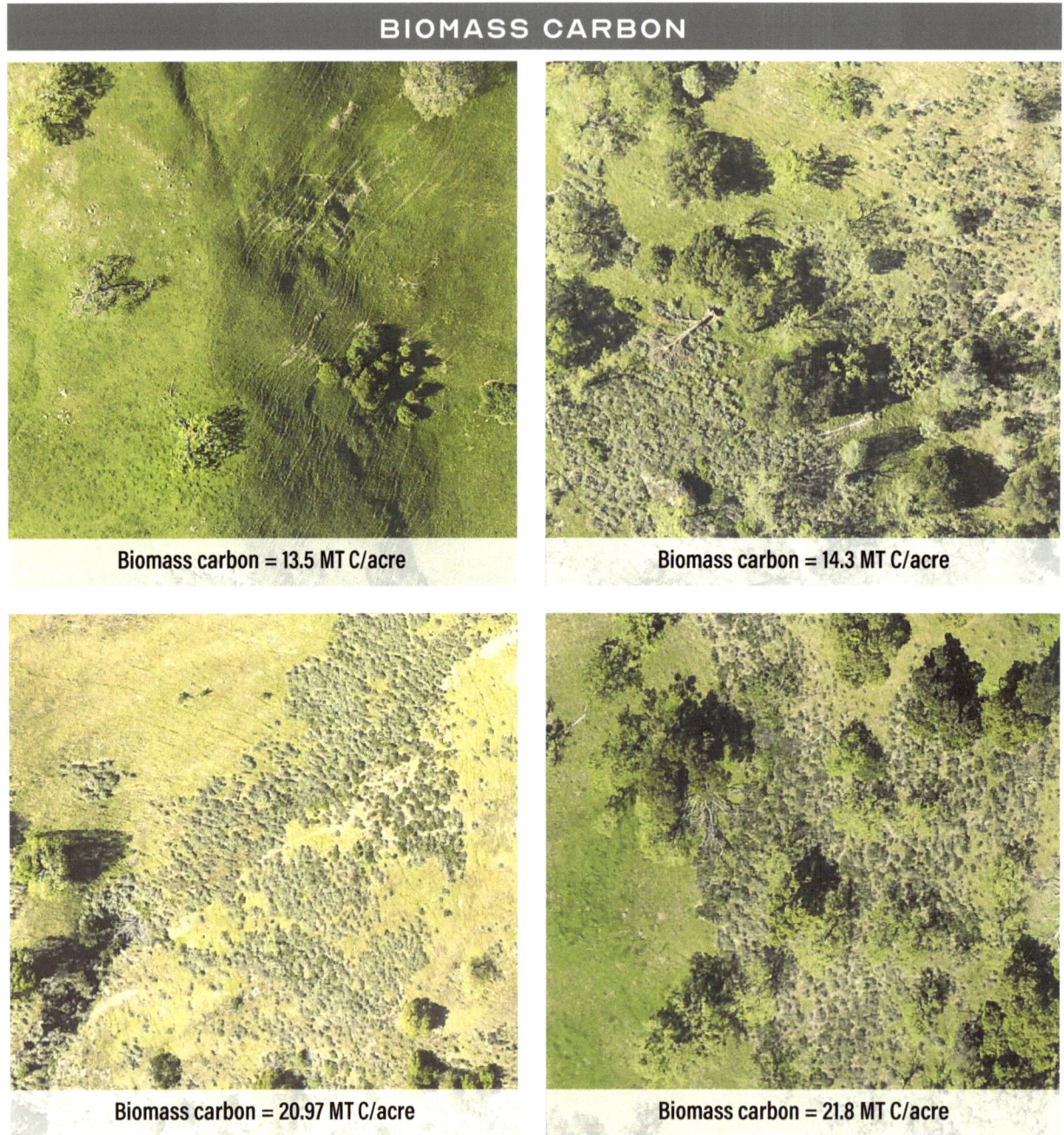

Biomass carbon = 13.5 MT C/acre

Biomass carbon = 14.3 MT C/acre

Biomass carbon = 20.97 MT C/acre

Biomass carbon = 21.8 MT C/acre

Figure 4.7. Examples of sites from the Alameda Watershed classified as savanna. Each image shown covers 9.9 acres (4 ha).

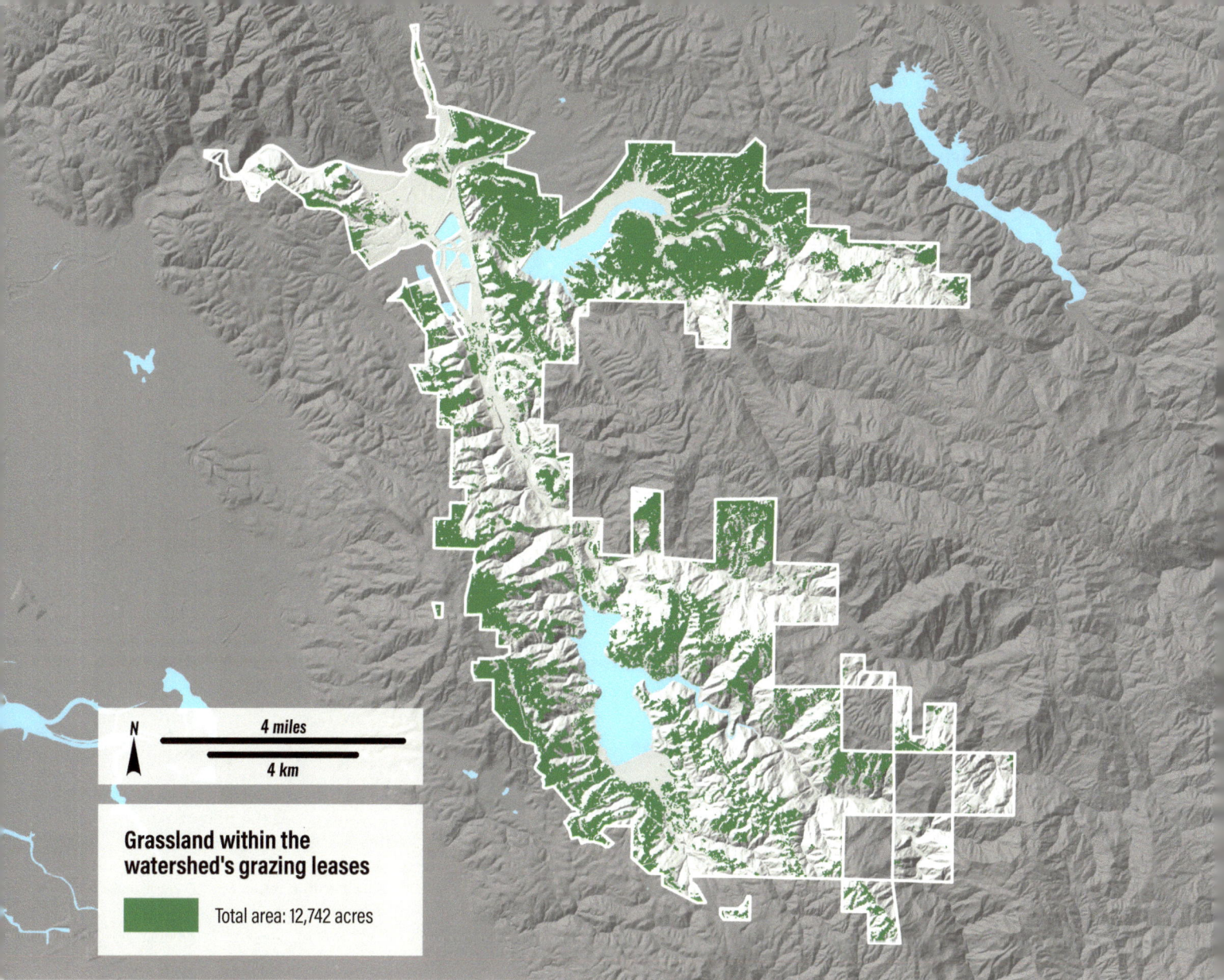

**Figure 4.8. A starting point for identifying potential opportunity space for silvopasture projects in the Alameda Watershed.** Sites represented in the mapping include all areas classified as grassland within the watershed's grazing leases. A number of other factors should be considered to identify sites where silvopasture may be an appropriate management strategy, such as historical ecology, land use history, soil characteristics, microclimate, depth to groundwater, and accessibility.

## SITE CONSIDERATIONS

The most appropriate sites for silvopasture within the Alameda Watershed are grazed grasslands with low existing tree cover, whose carbon sequestration rates may be increased through deliberate native tree plantings. Designs for silvopasture sites should strive to recreate the low stand density characteristic of historic oak savanna in the watershed (Stanford et al., 2013), rather than the thicker canopy cover associated with current oak woodland habitats. A relatively sparse canopy of oaks has the potential to support a mosaic of grassland and woody vegetation, increasing herbaceous biodiversity (Marañón and Bartolome, 1994) while still providing benefits of increased soil fertility (Camping, 2002; Dahlgren et al., 1997; Silver et al., 2010).

SILVOPASTURE

| CATEGORY | COSTS AND BENEFITS |
|---|---|

**AGRICULTURE**

**Co-benefits**

Silvopasture provides shade and shelter for livestock in rangelands, and in some cases can enhance forage production due to improved soil properties (Callaway et al., 1991; Dahlgren et al., 1997).

**Tradeoffs**

In some cases, the presence of oaks may inhibit grass productivity beneath trees (Callaway et al., 1991), impacting livestock forage production.

**NATIVE BIODIVERSITY**

**Co-benefits**

Native tree plantings associated with silvopasture applications provide habitat for woodland/savanna-dependent species, and a heterogeneous mix of oak canopy and open grassland can increase landscape-scale vegetation diversity (gamma diversity) due to differences in understory and open grassland communities (Stahlheber, 2016). Silvopasture also can facilitate the establishment of native grassland vegetation in the understory, as long as invasive species establishment and spread is controlled (Stahlheber and D'Antonio, 2014).

**Tradeoffs**

Canopy cover associated with silvopasture provides protection for predator species and offers perches for raptors, with potential negative impacts to other grassland wildlife and livestock. Enhanced soil fertility may also increase productivity of nonnative vegetation species (Stahlheber and D'Antonio, 2014). Additionally, planting trees as container stock from nurseries carries a risk of plant pathogen introductions (Frankel et al., 2020).

| CATEGORY | COSTS AND BENEFITS |
|---|---|

**SOIL QUALITY**

**Co-benefits**

Tree plantings associated with silvopasture increase soil fertility relative to nearby grassland (e.g., Camping, 2002; Dahlgren et al., 1997; Silver et al., 2010). Trees capture nutrients both belowground and in the canopy (Callaway et al., 1991; Perakis and Kellogg, 2007) while nourishing soils through leaf litter deposition (Dahlgren and Singer, 1991). Compared to open grasslands, soils beneath oaks are more productive (Waldrop and Firestone, 2006) and nutrient-rich (Camping, 2002; Carey et al., 2020; Dahlgren et al., 1997), less acidic (Camping, 2002; Dahlgren and Singer, 1991), and less prone to leaching and erosion-related losses (Dahlgren et al., 1997).

**WATER SUPPLY**

**Tradeoffs**

By increasing evapotranspiration, tree planting has the potential to reduce water supply within the watershed (Jackson et al., 2005). This tradeoff is expected to increase at higher planting densities.

## OTHER CONSIDERATIONS

The estimated carbon benefits of silvopasture provided above are based on average growth rates from a variety of sites and assume no tree mortality. In reality, actual growth rates of planted trees will depend on site conditions, and long-term survival of oak plantings may present a challenge for managers. An analysis of valley oak restoration sites near Vacaville, CA found that although emergence and survival rates can be very high with minimum management inputs (mulch and protection from herbivory, but no fertilization or irrigation), herbivory or other sources of mortality may limit long-time survival (Bernhardt and Swiecki, 2015). Model projections indicate that valley oaks in the region of the Alameda Watershed may be resilient to future warming and precipitation decreases (Kueppers et al., 2005), but climate change may have negative effects on growth rates of local populations (Browne et al., 2019). Additionally, increasingly common droughts and wildfires pose a challenge for seedling emergence and survival, particularly where populations are not locally adapted to future climate stressors (Browne et al., 2019; Mead et al., 2019). For silvopasture to provide long-term carbon and GHG benefits, these factors should be considered to determine appropriate seed sources and management practices favoring long-term survival of newly established trees. Even so, carbon sequestration in trees entails a risk of reversal due to wildfire, drought, or other disturbances that are expected to increase in frequency with climate change (Dass et al., 2018).

Hills, grasslands, and oaks surrounding reservoir, Alameda Watershed, courtesy of SFPUC.

# CATTLE EXCLUSION

## CARBON & CLIMATE BENEFITS

GHG benefits of cattle exclusion depend on the extent and rate of shrubland and woodland expansion.

Based on retrospective studies elsewhere in the Bay Area, excluding cattle would sequester between 0.08 and 0.21 MT C/acre per year in woody vegetation and soil over the coming decades.

Sequestering carbon and avoiding methane emissions from cattle would offset GHG emissions by 0.4 to 1 MT $CO_2$e/acre per year.

Excluding cattle from 80,000 to 200,000 acres of rangeland will offset ~1% of San Francisco's 1990 emissions.

## CO-BENEFITS

## TRADEOFFS

## ASSESSMENT

Grazing and wildfire pressure in grassland ecosystems effectively prevents the encroachment of shrubs and woody vegetation. The exclusion of grazers from parts of the Alameda Watershed has the potential to increase woody cover and associated aboveground biomass with implications for increased carbon storage in the watershed's large portion of leased rangeland. However, the carbon benefits of grazer exclusion must be weighed against the loss of grassland biodiversity support known to be provided by grazing activity and potential increased wildfire risk in shrublands.

In the absence of major land use modifications or regular disturbances like fire or grazing, vegetation communities in California have frequently been observed to transition from more open vegetation types, such as grasslands, to more closed vegetation mosaics, such as shrublands or woodlands (Keeley, 2005; Sandel et al., 2012). Numerous studies have documented conversion of grasslands to coastal scrub (often dominated by coyote brush [*Baccharis pilularis*]) or, less frequently, to chaparral (Callaway and Davis, 1993; Hobbs and Mooney, 1986; McBride and Heady, 1968; Russell and McBride, 2003; Williams et al., 1987). Coastal scrub and chaparral can facilitate a transition from grassland to oak woodland or mixed evergreen forest, and indeed evidence suggests that grassland conversion to oak-dominated hardwood forest generally transitions through an intermediate shrubland stage (Callaway and Davis, 1993; McBride, 1974; Mensing, 1998; Zavaleta and Kettley, 2006).

Disturbances can interrupt these successional pathways, and may be needed to maintain certain vegetation types. In many parts of the state, grasslands require periodic fire or grazing to prevent transition toward shrubland or woodland (Ford and Hayes, 2007; Tyler et al., 2007), except where other environmental conditions (such as serpentine soils) inhibit the transition to woody vegetation (Harrison and Viers, 2007). In the East Bay hills, for example, McBride (1974) found that grasslands exposed to cattle grazing had minimal coyote brush cover, while adjacent ungrazed grasslands experienced rapid coyote brush invasion.

Cattle grazing, Alameda Watershed, photograph by SFEI.

In the Alameda Watershed, livestock are an influential driver of vegetation patterns. Livestock grazing has occurred in the region since the early to mid-1800s, and approximately 31,000 acres within the watershed are currently grazed (SFPUC, 2017). The exclusion of livestock from currently grazed areas would thus likely result in expansion of coyote brush or other woody vegetation into existing grasslands in the watershed.

## CARBON AND GHG BENEFITS

The establishment or expansion of shrubs and trees that would be expected following grazer exclusion generally increases carbon storage in both aboveground and belowground pools, including shrub and tree biomass, leaf litter and downed wood, and soil carbon. For instance, in a study at Jasper Ridge Biological Preserve, Zavaleta and Kettley (2006) demonstrated how the encroachment of coyote brush into grasslands increased carbon storage. In the 25 years after coyote brush was established, aboveground carbon storage increased from 0.7 MT C/acre to 22.1 MT C/acre, with a total carbon increase of 36.7 MT C/acre. Sites continued to accumulate carbon 25 years after coyote brush was established at a rate of 1.5 MT C/acre-yr. Similarly, Daryanto et al. (2013) and Qiu et al. (2013) showed that grazer exclusion led to an increase in ecosystem carbon stocks, both aboveground and belowground.

The effect of cattle exclusion on ecosystem carbon stocks depends in large part on the trajectory of changes in the extent and density of woody vegetation. Retrospective studies using aerial photography have in some cases identified rapid shifts from grassland to shrubland after grazing pressures are released, as in the East Bay Regional Parks between 1939 and 1997 (Russell and McBride, 2003). In other cases, increases in woody vegetation after eliminating or reducing grazing have been less pronounced, as seen in Point Reyes between 1952 and 1993 (Russell and McBride, 2003). Vegetation changes may take place gradually over time, or may alternatively occur as rapid, episodic events that can occur years or decades after cattle are excluded and may be challenging to attribute conclusively to reduced grazing pressure (Williams et al. 1987).

Cattle grazing, Alameda Watershed, courtesy of SFPUC.

Observations from other open space sites within the Bay Area (Russell and McBride, 2003) or central coast (Callaway and Davis, 1993) have quantified changes in grassland, shrubland, and woodland or forest cover over periods of three to six decades following grazing elimination or reduction in grazing pressure. Carbon storage estimates from the Alameda Watershed were applied to these observed transition rates to estimate potential effects of cattle exclusion on vegetation carbon storage in grassland, coastal scrub and chaparral, and oak savanna, oak woodland, and riparian forest. Where the extent of woody vegetation increased, a mean annual soil carbon accumulation rate of 0.3 MT C/acre-yr was used to estimate soil carbon sequestration, based on observations under coyote brush from Zavaleta and Kettley (2006). Resulting mean annual carbon sequestration rates ranged from 0.08 to 0.21 MT C/acre-yr. While this range provides a reasonable approximation for how cattle exclusion may affect Alameda Watershed carbon stocks, actual changes in the vegetation mosaic will depend in part on specific site conditions and other disturbances, which may interact with the elimination of grazing pressure. Among seven sites evaluated by Russell and McBride (2003), for example, those with the greatest initial grassland cover generally saw the largest increase in woody vegetation.

In addition to influencing the vegetation mosaic, cattle affect the climate through the emission of methane produced by enteric fermentation. Based on IPCC per-head methane emission factors (IPCC, 2006) and stocking rates from the draft Alameda Creek Watershed Rangeland Management Plan (SFPUC, 2017), elimination of methane emissions from cattle would provide an additional GHG benefit of 0.09 to 0.27 MT $CO_2$e/acre-yr. Combining these avoided cattle emissions and vegetation carbon sequestration, cattle exclusion offers an estimated GHG benefit of 0.4 to 1 MT $CO_2$e/acre-yr. This estimate assumes that excluding cattle from the watershed would not increase the extent of grazed land or cattle stocking rates elsewhere. Any such increases in the number of cattle elsewhere would negate the benefit of avoided methane emissions.

| CATEGORY | COSTS AND BENEFITS |
|---|---|

**NATIVE BIODIVERSITY**

**Co-benefits**

Shrub cover within grassland or chaparral ecosystems may provide refuge from predation and facilitate movement for native species such as California tiger salamander or California red-legged frog (*Rana aurora draytonii*) (Bulger et al., 2003; Wang et al., 2009).

**Tradeoffs**

A variety of flora and fauna in native grasslands are supported by grazing activity. Cattle exclusion and subsequent woody encroachment may alter the distribution of native wildlife in such habitats. Evidence indicates that a variety of grassland fauna, including checkerspot butterflies (*Euphydryas spp.*) (Weiss, 1999), grassland songbirds (Gennet et al., 2017), burrowing owls (Haug and Oliphant, 1990), California tiger salamanders, and California red-legged frogs (Bartolome et al., 2014), are supported by grazing activities in grassland ecosystems, which include the maintenance of stock ponds and grassland habitat for a variety of wildlife. Grazing also supports the native flora and fauna of serpentine grasslands (Bartolome et al., 2014). The presence of grazing livestock has also been seen to reduce the prevalence of noxious weeds (Malmstrom et al., 2017).

**WILDFIRE**

**Tradeoffs**

The increase in aboveground biomass associated with woody encroachment can increase the likelihood of high-intensity wildfires and extreme fire behavior that can be particularly challenging to control (Parsons et al., 2016; Russell and McBride, 2003). As wildfires represent a source of carbon and other climate pollutants to the atmosphere in the short-term, this risk must be weighed with the carbon benefits of increased woody vegetation.

| CATEGORY | COSTS AND BENEFITS |
|---|---|

**WATER QUALITY**

**Co-benefits**

Cattle exclusion eliminates a potential source of pathogens, nutrients, and physical degradation to riparian corridors, wetlands, and channels (Herbst and Knapp, 1995). This benefit may be minimal on the Alameda Watershed, given that grazing leases are required to conform to best management practices that reduce risks of pathogen contamination and other negative water quality effects (SFPUC, 2017), such as providing adequate buffers around water bodies to minimize pathogen risk (Tate et al., 2006).

**AGRICULTURE**

**Tradeoffs**

Cattle exclusion reduces the amount of land available for grazing on existing leases, with economic and financial implications that also include increased maintenance time for fencing.

**CATTLE EXCLUSION**

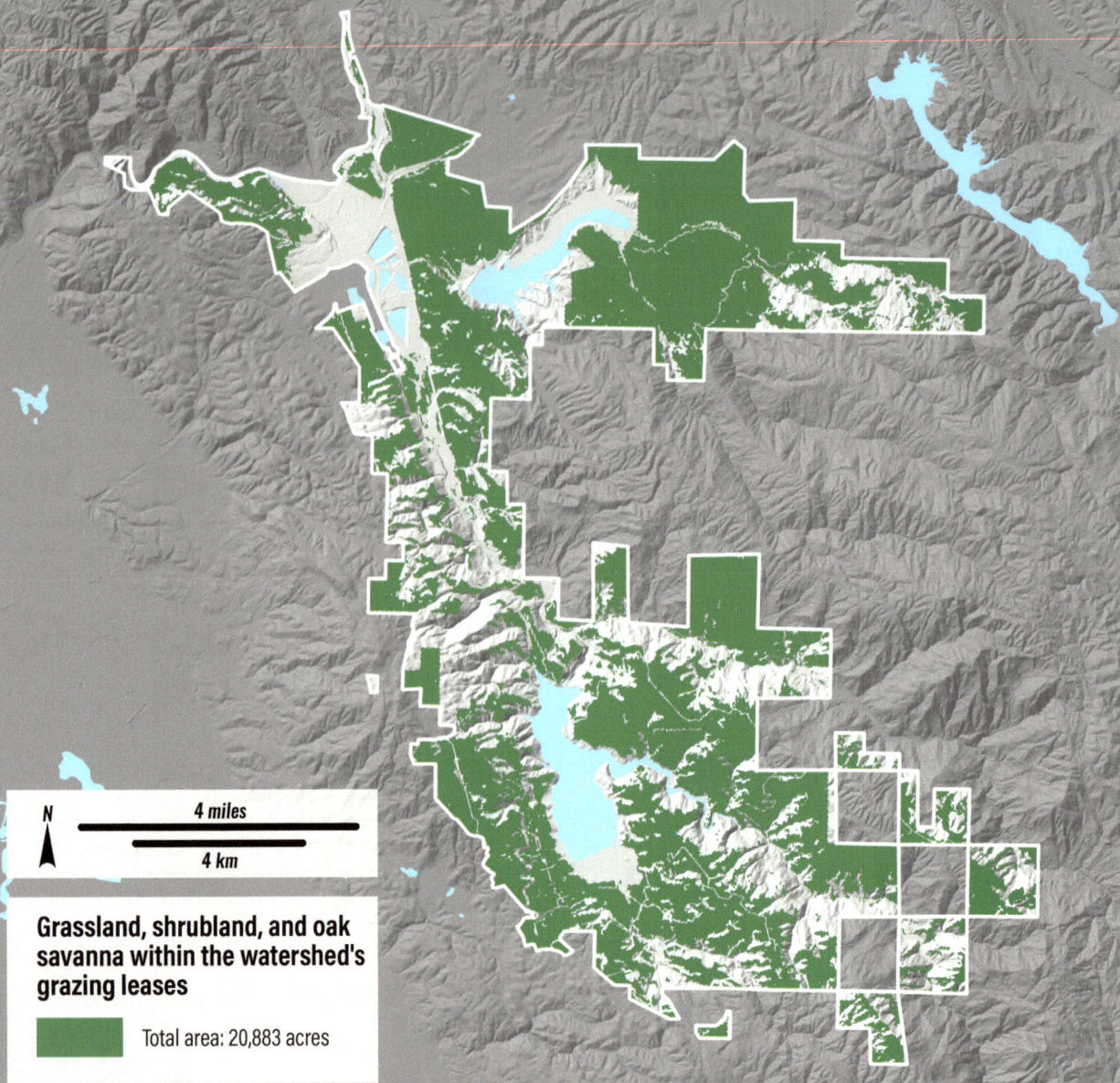

N

4 miles

4 km

**Grassland, shrubland, and oak savanna within the watershed's grazing leases**

Total area: 20,883 acres

**Figure 4.10. A starting place for identifying the potential opportunity space for cattle exclusion projects in the Alameda Watershed.** Sites represented in the mapping include all vegetated areas not classified as oak woodland or riparian forest within the watershed's grazing leases. In addition to vegetation type and existing grazing leases, a number of other factors should be considered to identify sites where cattle exclusion might be an appropriate management strategy, such as grazing intensity and timing, historical ecology, native plant cover, presence of serpentine substrate, presence of rare and endangered plants and animals, and modeled effects on wildfire and vegetation succession. As noted above, there are a number of major concerns associated with cattle exclusion, and the mapping does not assess the desirability of applying this strategy within potential opportunity areas.

## SITE CONSIDERATIONS

The maximum potential area in which a grazing exclusion management strategy could apply includes all portions of the watershed that fall within current grazing leases outside of wood-land habitats[3] (Fig. 4.10). Within this footprint, grazing exclusion may be most appropriate in areas that were historically dominated by woody vegetation, or in areas currently dominated by coastal scrub, chaparral, or savanna where an increase in woody biomass is desirable. The effect of cattle exclusion on ecosystem carbon stocks may be greatest in areas that are most actively

[3] We assume that grazing exclusion from existing woodlands would result in no further increase in carbon storage.

utilized by livestock. However, there are likely to be substantial tradeoffs to grazing exclusion in certain settings, and thus this strategy may not be compatible with SFPUC management objectives in many places in the watershed.

Grazing by cattle and other livestock can have a strong influence on grassland vegetation communities. In some cases, grazing has been found to benefit native vegetation. Studies from central California, for example, have found that livestock grazing can benefit native forbs (reviewed in Bartolome et al., 2014). However, this relationship is variable across sites, plant guilds, and studies (e.g., Gornish et al., 2018; Hayes and Holl, 2003; Holl and Hayes, 2006; Mariotte et al., 2017), and grazing is generally understood to have less of an effect on native and invasive grasses (Bartolome et al., 2014; Hayes and Holl, 2003). In serpentine sites, livestock grazing can benefit native serpentine grassland communities. Several studies from central California have found that grazing generally benefits native plants and decreases nonnative cover in serpentine areas, particularly in the presence of background nitrogen deposition (Beck et al., 2015; Funk et al., 2015; Harrison et al., 2003; Pasari et al., 2014). The effect of grazing on vegetation communities may depend on the timing of grazing and other factors. Stahlheber and D'Antonio (2013), for example, found that winter and early spring grazing was most beneficial for native grassland species. Similarly, targeted grazing can be an effective strategy for managing invasive weeds (DiTomaso, 2000; Malmstrom et al., 2017). Elimination of cattle from the watershed's grasslands has the potential to alter the cover and diversity of native and nonnative vegetation.

Grassland wildlife may also be influenced by grazing exclusion. Cattle grazing on California rangelands can maintain foraging and breeding habitat for grassland birds, a guild of birds that are in steep decline across western North America. Moderate levels of livestock grazing have been linked to native songbird conservation through the positive effects of grazing on native vegetation cover and structural heterogeneity (Gennet et al., 2017), and burrowing owls require the short vegetation and matrix of open sites maintained by grazing livestock (Haug and Oliphaunt, 1990). Cattle grazing is generally considered beneficial for California's red-legged frog and tiger salamander, as it provides breeding sites in stock ponds and maintains areas of grassland (Bartolome et al., 2014). Additionally, grazing may increase the abundance of California ground squirrels (*Otospermophilus beecheyi*), though this relationship is unclear (Bartolome et al., 2014; Fehmi et al., 2005). However, grazing activities also increase competition for wildlife that rely on foraging (Fehmi et al., 2005). Due to grazing's nuanced role in ecosystem biodiversity, grazing exclusion should be applied strategically throughout a landscape, as the carbon benefits of saturating a habitat with woody vegetation may be outweighed by negative ecological impacts on vegetation and wildlife communities. The use of pilot studies with comprehensive vegetation, wildlife, and carbon monitoring would offer site specific information on the effects of cattle exclusion on SFPUC management objectives.

Fire also plays a significant role in the trajectory of vegetation communities, and the watershed's vulnerability to high-severity wildfire will increase if fuel loads are no longer controlled by grazing cattle. The carbon benefits provided by grazer exclusion in the watershed would need to be weighed against the increased risk of wildfire, which, among other negative outcomes, triggers a short-term release of carbon (see Chapter 3) and emits other potent climate pollutants such as methane and black carbon.

# NATIVE GRASSLAND RESTORATION

## CARBON & CLIMATE BENEFITS

Restoring native grassland vegetation sequesters 0.093 MT C/acre per year in degraded or sparsely vegetated sites. (Carbon and GHG benefits of restoring invaded grasslands to native vegetation are not well understood.)

Offsets GHG emissions by 0.34 MT $CO_2$e/acre per year in degraded or sparsely vegetated sites.

Reseeding 200,000 acres of bare or sparsely vegetated land could offset an estimated 1% of San Francisco's 1990 emissions.

## CO-BENEFITS

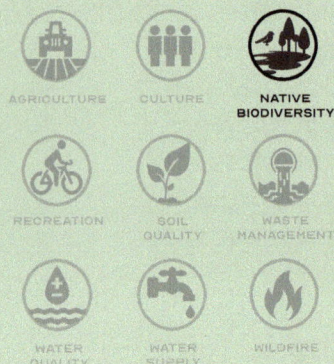

## TRADEOFFS

## ASSESSMENT

Restoring native grassland vegetation can sequester carbon, enhance biodiversity, and support soil health in degraded or unvegetated sites. Elsewhere in the watershed, the potential for native grassland restoration to halt or reverse soil carbon losses is not well understood.

Grasslands are among the most highly altered of California's ecosystems. In many areas, introduced nonnative annual grasses or forbs have largely displaced native grassland species, with major implications for biodiversity support, disturbance regimes, and nutrient cycling (D'Antonio et al., 2007). Further loss and degradation of native grasslands has occurred through urban and agricultural development, overgrazing, nitrogen deposition, and encroachment of woody vegetation resulting from fire suppression and other changes in disturbance regimes (Huntsinger et al., 2007; Keeley, 2005; Pasari et al., 2014). Nonnative annual grassland is currently the most extensive vegetation type within the Alameda Watershed, occupying 20,614 acres, and is dominated by species such as wild oat (*Avena spp.*), wild barley (*Hordeum spp.*), and brome grasses (*Bromus spp.*) (SFPUC, 2015). Small patches of native valley needlegrass grassland and serpentine bunchgrass grassland persist in some locations, and other sites support high native species richness and a number of locally rare plant species. Many areas within the watershed that are dominated by nonnative vegetation also support high biodiversity of native herbaceous plants, including special-status plant species and host plants for special-status wildlife (ACRCD and LD Ford, 2018b). Native grasslands likely existed historically in many of the areas occupied by nonnative annual grasslands today, though the pre-colonization composition of these grasslands is not well understood; perennial bunchgrasses such as purple needlegrass (*Stipa pulchra*) likely dominated in some areas, while perennial or annual forbs dominated in others (Evett and Bartolome, 2013; Stanford et al., 2013).

The displacement of native perennial grasslands by nonnative annual grasslands has likely resulted in a decrease in soil carbon storage. Native perennial grasses begin regrowing early in the fall and can grow longer into the summer than the annual grasses common to California grasslands, resulting in greater annual net primary productivity (Eviner, 2016). In addition, native perennial grasses have much deeper root systems than annual grasses, and thus sequester carbon at greater depths in the soil (DuPont et al., 2010; Wilsey and Wayne Polley, 2006). Koteen et al. (2011), for instance, found that grasslands in Marin County have lost an average of 16 MT/acre of soil carbon in the top 50 cm over the past ~200 years due to conversion to nonnative annual grassland. Such findings may not apply to invaded grasslands that were historically dominated by annual forbs, which are thought to have been common throughout California (Evett and Bartolome, 2013).

By most estimates, annual grasslands in California represent a net carbon source, but the magnitude is not well constrained. Recent values reported for California grasslands range widely among years, sites, and quantification methods (Chou et al., 2008; Ma et al., 2007; Mayer and Silver, 2022; Owen et al., 2015; Ryals et al., 2015; Xu and Baldocchi, 2004), from net sequestration of 0.35 MT C/acre from 2004–2005 in a foothill grassland (Xu and Baldocchi, 2004) to a net carbon loss of 0.92 MT C/acre at a nearby site between 2008 and 2011 (Ryals et al., 2015). In contrast, loss of native grasslands due to woody encroachment is likely associated with an increase in ecosystem carbon storage due to increased aboveground biomass and soil organic carbon (Eve et al., 2014; Zavaleta and Kettley, 2006).

## CARBON MANAGEMENT POTENTIAL

While the conversion of native perennial grassland to nonnative annual grassland is thought to decrease ecosystem carbon storage, the carbon benefits of native grassland restoration have not been conclusively demonstrated. The Comet Planner tool used by the California Department of Food and Agriculture (CDFA) Healthy Soils Program (HSP) estimates a carbon sequestration benefit of 0.34 metric tons $CO_2$ equivalent per acre per year for restoring degraded rangelands with limited plant cover (through reseeding with either native or nonnative species) in Alameda County (http://bfuels.nrel.colostate.edu/health#). However, land use history can have long lasting effects on soil microbial communities and carbon cycling, which can persist for multiple years post-restoration; in previously cultivated areas, restoration of a native grassland community therefore may not be accompanied by simultaneous re-establishment of the pre-modification microbial community or soil carbon processes (Jackson et al., 2007). In addition, restoration approaches that use tillage and herbicide application prior to seeding with native grasses (to deplete the nonnative annual seed bank) may result in a temporary net loss of total soil carbon from the upper soil layers (<15 cm depth; Potthoff et al., 2005; Steenwerth et al., 2006), though this effect would likely subside over time. Thus, while native grassland restoration could be expected to slow or halt ongoing carbon losses associated with nonnative annual grasslands (see above), further research is needed to better understand the time horizon and the ultimate magnitude of carbon accumulation associated with restoration.

In addition to uncertainty regarding both short- and long-term carbon sequestration potential, there are a number of other considerations associated with grassland restoration that must be taken into account when evaluating its feasibility and effectiveness as a management strategy. Restoring native grasslands can benefit a wide range of native plants, insects, birds, and other taxa, including special status species such as the Western burrowing owl (*Athene cunicularia* ssp. *hypugaea*; Artis, 2011; Luong et al., 2019; Suttle and Thomsen, 2007). Native grassland vegetation has been shown to be more effective at suppressing noxious weeds than naturalized exotic species (Eviner and Malmstrom, 2018). Additionally, models suggest that, as a carbon sink, grasslands in California may be more resilient to climate change than forests,

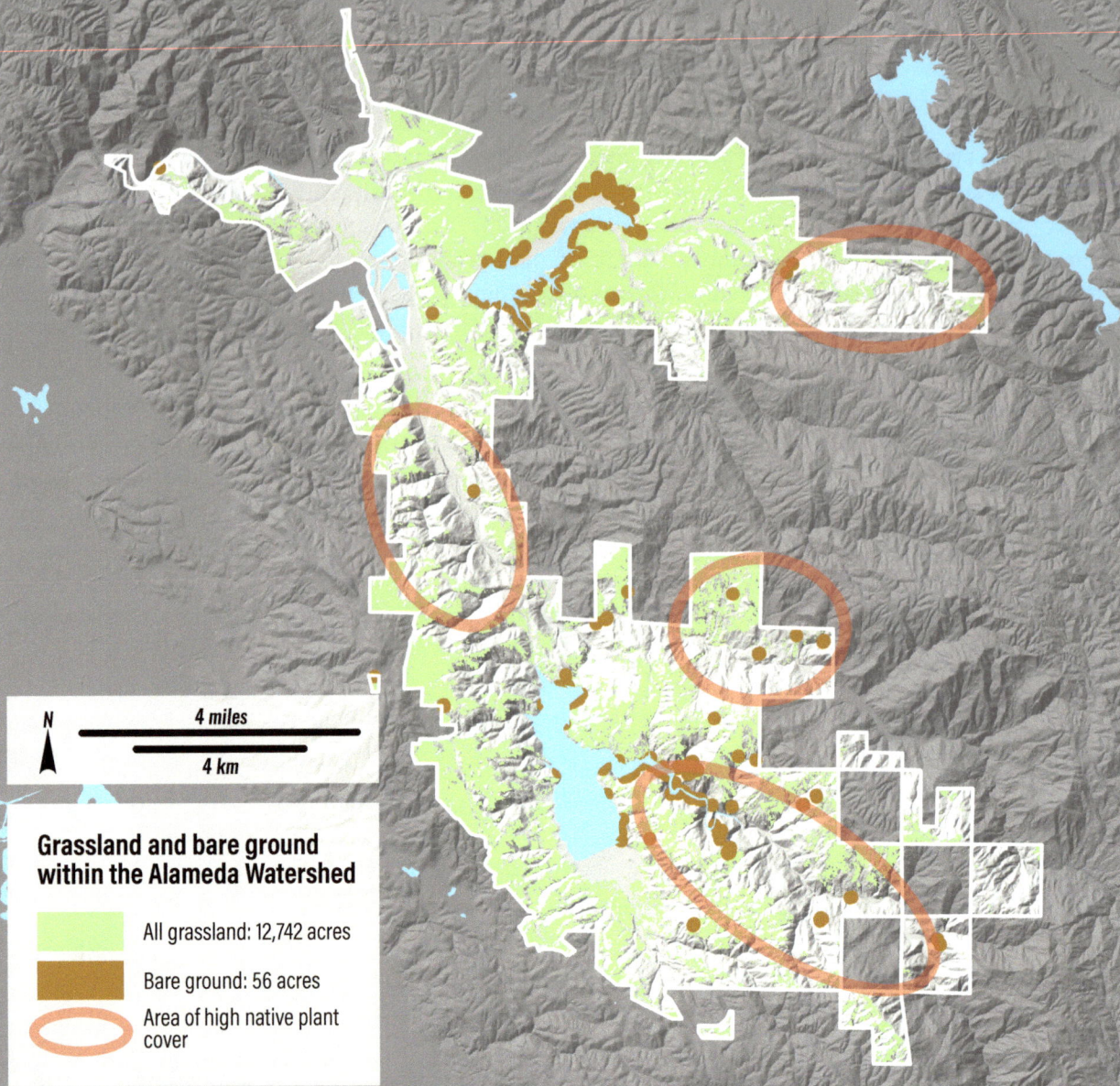

**Figure 4.11. Potential opportunity space for native grassland restoration projects.** Potential opportunity space for native grassland restoration projects. Opportunity space includes areas classified as bare ground (brown) or grassland (green). Circles indicate general areas where high native plant cover has been reported.

where carbon stocks are more susceptible to drought and wildfire (Dass et al., 2018). However, past efforts to restore native grasslands have met with mixed success. Successful restoration of native vegetation is highly labor and time intensive, requiring extensive long-term management (Stromberg et al., 2007). The effectiveness of grassland restoration depends on many factors, including legacy effects from past land uses, prevalence of invasive species, planting and site preparation techniques, invasive species control techniques (e.g., herbicide application, prescribed burning, biological control), soil nitrogen availability, and rainfall regime, among others (Buisson et al., 2008; Corbin et al., 2004; Nolan et al., 2021; Stromberg et al., 2007; Suttle and Thomsen, 2007).

## SITE CONSIDERATIONS

Within the Alameda Watershed, existing nonnative annual grasslands with some native vegetation cover, or that are adjacent to sites with high native grassland vegetation cover, are likely to be the most suitable sites for native grassland restoration. Such sites may offer suitable conditions for native grassland species as well as lower competition from nonnative annuals than more heavily invaded sites (Gornish and Ambrozio dos Santos, 2016). Recent invasive plant inventories (Nomad Ecology, 2020) and rangeland monitoring (ACRCD and LD Ford, 2018b) in the watershed provide baseline information regarding invasive and native plant cover and presence of sensitive grassland habitats, which can be used to help prioritize locations for grassland restoration along with other management strategies (Fig. 4.11). Sites located along Upper Alameda Creek, Upper San Antonio Creek, and to the west of Sunol Valley with low to moderate invasive plant cover represent potentially high priority sites for considering native grassland restoration. Restoration of native vegetation in these sites may limit or potentially reverse ongoing soil carbon losses, but this effect has not been observed in field trials. Degraded sites with bare ground or sparse vegetation cover are also potentially high priority for reseeding with native grassland species. In such sites, reestablishment of grassland vegetation is expected to increase soil carbon storage and improve soil health (Swan et al., 2015). While such sites are present on the Alameda Watershed, their overall coverage is limited. Excluding areas bordering the reservoirs, only 27 acres within the watershed are classified by LANDFIRE data as bare ground or sparsely vegetated.

**CATEGORY**

**NATIVE BIODIVERSITY**

**COSTS AND BENEFITS**

**Co-benefits**

Native grasslands support a wide range of native grasses and forbs, provide habitat for many native insects, birds, and other wildlife, and can more effectively suppress noxious weeds than grasslands dominated by naturalized exotic species (Eviner and Malmstrom, 2018).

**Tradeoffs**

The use of container stock from nurseries for revegetation projects carries a risk of plant pathogen introductions (Frankel et al., 2020)

Grasslands in the Alameda Watershed, photo by SFEI.

# OPEN SPACE CONSERVATION

## CARBON & CLIMATE BENEFITS

Conservation of the Alameda Watershed has avoided carbon losses ranging from ~300,000 MT C under urban development up to ~1,000,000 MT C under agricultural cultivation, equal to 1-4 MMT $CO_2$.

Per-acre carbon savings are highest in areas of riparian forest (60-90 MT C/acre), and lowest in open grassland (-5-20 MT C/acre).

Additional GHG emissions from fossil fuels and fertilizer equal ~2-6 MT $CO_2$e/acre per year from agriculture and ~20-30 MT $CO_2$e/acre per year from nearby urban areas.

Protecting 700-2,000 additional acres of land[4] from urban or cropland development would avoid a loss of ecosystem carbon comparable to 1% of San Francisco's 1990 emissions. This value does not include fossil fuel or fertilizer emissions from land conversion, urban land use, or agriculture.

## CO-BENEFITS

AGRICULTURE  CULTURE  NATIVE BIODIVERSITY

RECREATION  SOIL QUALITY  WASTE MANAGEMENT

WATER QUALITY  WATER SUPPLY  WILDFIRE

## TRADEOFFS

CULTURE

## ASSESSMENT

In addition to maintaining a clean water supply and supporting grassland and woodland biodiversity, conservation of the Alameda Watershed as protected open space has avoided potential carbon losses from biomass and soil due to urban and agricultural development.

While the Alameda Creek watershed is not under threat from urban or agricultural development, calculating the carbon benefit derived from existing open space conservation in the watershed provides important context for interpreting the potential carbon and GHG benefits of other management strategies. The watershed represents a large, contiguous open space within a highly developed region with a growing population, and the existing carbon storage benefits of the watershed contrast notably with many surrounding areas. This section considers the avoided carbon losses attributable to continued open space conservation, relative to counterfactual scenarios in which the watershed is converted to urban or agricultural development.

## CARBON AND GHG BENEFITS

Urbanization of previously undeveloped land can increase or decrease carbon stocks depending on historical land cover, urban land uses and morphology, regional climate, and other factors. (Such changes in carbon stocks pertain to carbon presently stored in vegetation and soil, and do not account for other carbon impacts of urbanization such as construction, materials production, transportation, building energy use, and other sources of fossil fuel emissions. These additional effects are briefly discussed below.) With respect to carbon stocks, researchers have posited the idea of "urban convergence," suggesting that urbanization will tend to increase carbon storage in arid or semiarid regions with relatively low biomass, and decrease carbon storage in temperate regions with relatively high biomass (Pouyat et al., 2006). Evidence for this phenomenon in semiarid and arid regions is mixed, however. For instance, an analysis employing space-for-time substitution found that carbon storage in semiarid cities such as Oakland and Sacramento

---

[4] This acreage assumes that the land protected from development has a vegetation mosaic similar to the Alameda Watershed.

was greater than surrounding Mediterranean shrublands, while carbon storage in a more arid city (Phoenix, AZ) was lower than surrounding desert shrublands (McHale et al., 2017). In Silicon Valley, Beller et al. (2020) found that areas that historically supported oak woodlands had lost carbon as a result of urban development, while non-forested areas had generally gained carbon (assuming a carbon storage value of 0 for non-forested land cover types), resulting in a more uniform distribution of carbon today. Overall estimates for changes in tree carbon storage resulting from urbanization in Silicon Valley range from a non-significant gain of ~14% to a significant loss of ~60% (Beller et al., 2020). Values for urban carbon storage in trees and shrubs were used to estimate avoided carbon losses from urban development for ecosystems in the Alameda Creek watershed (Table 4.1). Tree carbon storage was taken from Beller et al. (2020), and urban shrub carbon was estimated as 5% of urban tree carbon, based on observations from both arid and humid cities (McHale et al., 2017; Nowak, 1994). Dead wood and litter in urban sites was assumed to be removed from the site and chipped, a common practice for urban tree and landscape residue (Whittier et al., 1994).

The contemporary distribution of carbon in urban areas also depends on land use and % impervious cover. Carbon storage tends to be higher in residential areas or areas with lower impervious surface cover than in commercial and industrial areas or along transportation corridors, which tend to have a greater percentage of impervious cover. This is a function both of greater tree cover and aboveground biomass in residential areas (Beller et al., 2020; Hutyra et al., 2011; McHale et al., 2017), as well as the higher levels of soil organic carbon associated with lawns and other pervious surfaces compared with impervious land cover types (Pouyat et al., 2006; Raciti et al., 2012; Yan et al., 2015). For instance, Raciti et al. (2012) found that soils in New York City covered by impervious surfaces had 66% lower carbon content than nearby open areas (e.g., lawns and median strips), and suggested that carbon losses from soils beneath impervious surfaces could be due to a combination of decomposition, aqueous losses, and topsoil removal from construction activities. Overall, however, the long-term effects of urban development on soil carbon have been observed to vary widely across cities (Pouyat et al., 2006), making it difficult to predict how urbanization would affect carbon storage in the Alameda Watershed's soils.

In addition to changes in ecosystem carbon storage, new development entails substantial GHG emissions due to activities such as construction, road-building, earth moving, and materials production and transport (EPA, 2009). Urbanization also leads to increased ongoing anthropogenic

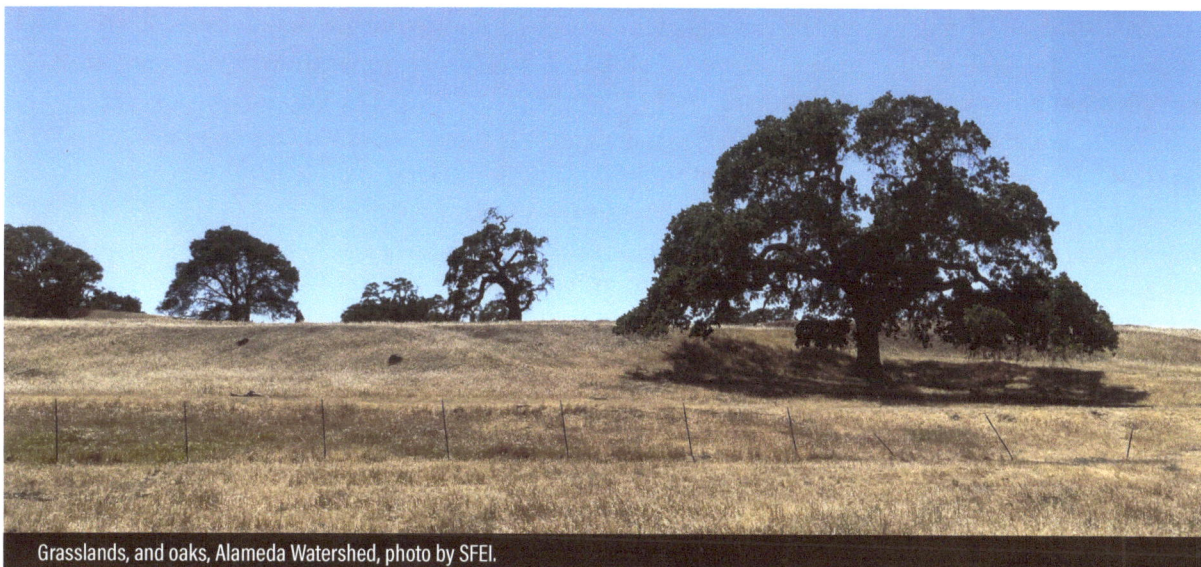

Grasslands, and oaks, Alameda Watershed, photo by SFEI.

GHG emissions from transportation, building energy use, wastewater management, turf maintenance, and other sources, which need to be considered for a full accounting of the benefits of open space conservation (Golubiewski, 2006). The City of Fremont, for instance, reported annual emissions of ~20–21 MT $CO_2$e/acre in 2020 (https://www.fremont.gov/about/sustainability/climate-action-plan-update), while the City of Livermore reported annual emissions of ~34 MT $CO_2$e/acre in 2017 (Rincon Consultants, Inc., 2020). Even urban land cover types that sequester carbon may become net GHG sources as a result of management activities; a study by Townsend-Small and Czimczik (2010) in Irvine, CA found that, while some urban lawns sequestered significant amounts of carbon, GHG emissions associated with turf maintenance (irrigation, fertilizer production, and fuel use) greatly exceeded the GHG benefits of that carbon storage.

As with urbanization, conversion of natural ecosystems to agricultural use generally results in a decrease in both biomass and soil carbon storage (Albaladejo et al., 2013; Boix-Fayos et al., 2009; Schlesinger, 1986). The magnitude of the effect depends largely on the historical land cover and the type of agriculture. In comparison with natural ecosystems, annual croplands generally have low biomass carbon storage; for temperate annual croplands, the IPCC provides a default value of 2 MT C/acre (IPCC 2006). Belowground, soil carbon decreases by ~20–40% with conversion of forest to cropland (global averages; Guo and Gifford, 2002; Murty et al., 2002). In the Alameda Creek watershed, conversion to annual crops would lead to an estimated loss of carbon in all ecosystems, with an average loss of 14 MT C/acre from biomass and 12–24 MT C/acre from soil (Table 4.1).

Perennial crops, and in particular woody crops such as vineyards, can regenerate a standing stock of biomass carbon over time, though carbon stocks generally do not recover to pre-agricultural levels (Williams et al., 2011). Smart (2003) found that former oak woodlands and oak savanna in Napa Valley that were converted to vineyards lost about 13 MT/acre of soil carbon in the top 30 cm (Smart et al., 2003); this is consistent with findings from Mendocino County, where vineyards were found to have 12–15% less carbon in the top meter of soil than adjacent wildlands (Williams et al., 2011). In the Alameda Creek watershed, conversion to vineyards would lead to an estimated loss of biomass carbon (in both the short term and the long term, after vines had developed) in all ecosystems except grassland, with an average loss of 12 MT C/acre; soil carbon would decrease by an estimated 7–9 MT C/acre on average (see Table 4.1).

As with urbanization, management practices associated with conversion of natural lands to agriculture alter net GHG emissions in a number of other ways. Nitrogen application through fertilizers or other amendments can result in substantial $N_2O$ emissions. A review by Verhoeven et al. (2017) across multiple annual and perennial crop types in California found annual $N_2O$ emissions ranging from 0.14 to 1.9 MT $CO_2$e/acre-yr. For annual crop systems, Verhoeven et al. (2017) found emissions of 0.33 ± 0.17 MT $CO_2$e/acre-yr, while a model-based analysis by De Gryze et al. (2010) estimated emissions of 0.45 ± 0.063 MT $CO_2$e/acre-yr for corn, cotton, sunflower, and wheat. For vineyards, Verhoeven et al. (2017) reports $N_2O$ emissions of 0.34 ±0.24 MT $CO_2$e/acre-yr. Accounting for other life-cycle emissions such as on-farm machinery, irrigation, trucking and storage, and fertilizer production, other studies have estimated the total GHG footprint of CA crop production to range from 0.5 MT $CO_2$e/acre-yr for organic alfalfa to as high as 7 MT CO2e/acre-yr for conventionally-grown berry crops (Shaffer and Thompson, 2015; Venkat, 2012). In comparison, the cattle raised on Alameda Watershed grazing leases emit an estimated 0.1–0.3 MT $CO_2$e/acre-yr of $CH_4$ due to enteric fermentation, based on IPCC default per-head emission factors (IPCC, 2006) and recommended stocking rates in the draft Alameda Creek Watershed Rangeland Management Plan (SFPUC, 2017).

| CATEGORY | COSTS AND BENEFITS |
|---|---|

**AGRICULTURE**

**Co-benefits**

Open space conservation enables portions of the Alameda Creek watershed to be used for grazing land.

**CULTURE**

**Co-benefits**

Open space conservation supports the region's agricultural heritage and economy.

**Tradeoffs**

Constraints on development within the watershed may increase the regional cost of living.

**NATIVE BIODIVERSITY**

**Co-benefits**

The large contiguous areas of open space protected in the Alameda Creek watershed provide extensive habitat and refugia for a wide range of wildlife, and play a major role in the conservation of locally rare plant and animal species and vegetation communities. The watershed is designated as an area "essential" to regional conservation goals by the Conservation Lands Network (Bay Area Open Space Council, 2019). The watershed's protected lands provide opportunities for wildlife movement where Alameda Creek passes under Highway 680, and enhance landscape resilience by providing important connectivity between adjacent open spaces in the East Bay Hills and Hamilton Range.

| CATEGORY | COSTS AND BENEFITS |
|---|---|

**RECREATION**

**Co-benefits**

Portions of the watershed within Sunol Regional Park are used for hiking, horseback-riding, nature access, and other recreational uses.

**SOIL QUALITY**

**Co-benefits**

Conserving natural ecosystems can avoid soil degradation due to topsoil removal and other construction-related disturbances, the use of impervious surfaces, and agricultural tillage.

**WATER QUALITY**

**Co-benefits**

Conserved open space helps to protect water quality by limiting contaminant sources and maintaining permeable surfaces with high infiltration.

**WATER SUPPLY**

**Co-benefits**

Compared with impermeable surfaces, natural ecosystems have greater water infiltration and reduced runoff (Arnold and Gibbons, 1996). By reducing runoff volume and releasing water more slowly, conserving open space in reservoir catchments reduces the need for greater reservoir capacity.

OPEN SPACE CONSERVATION

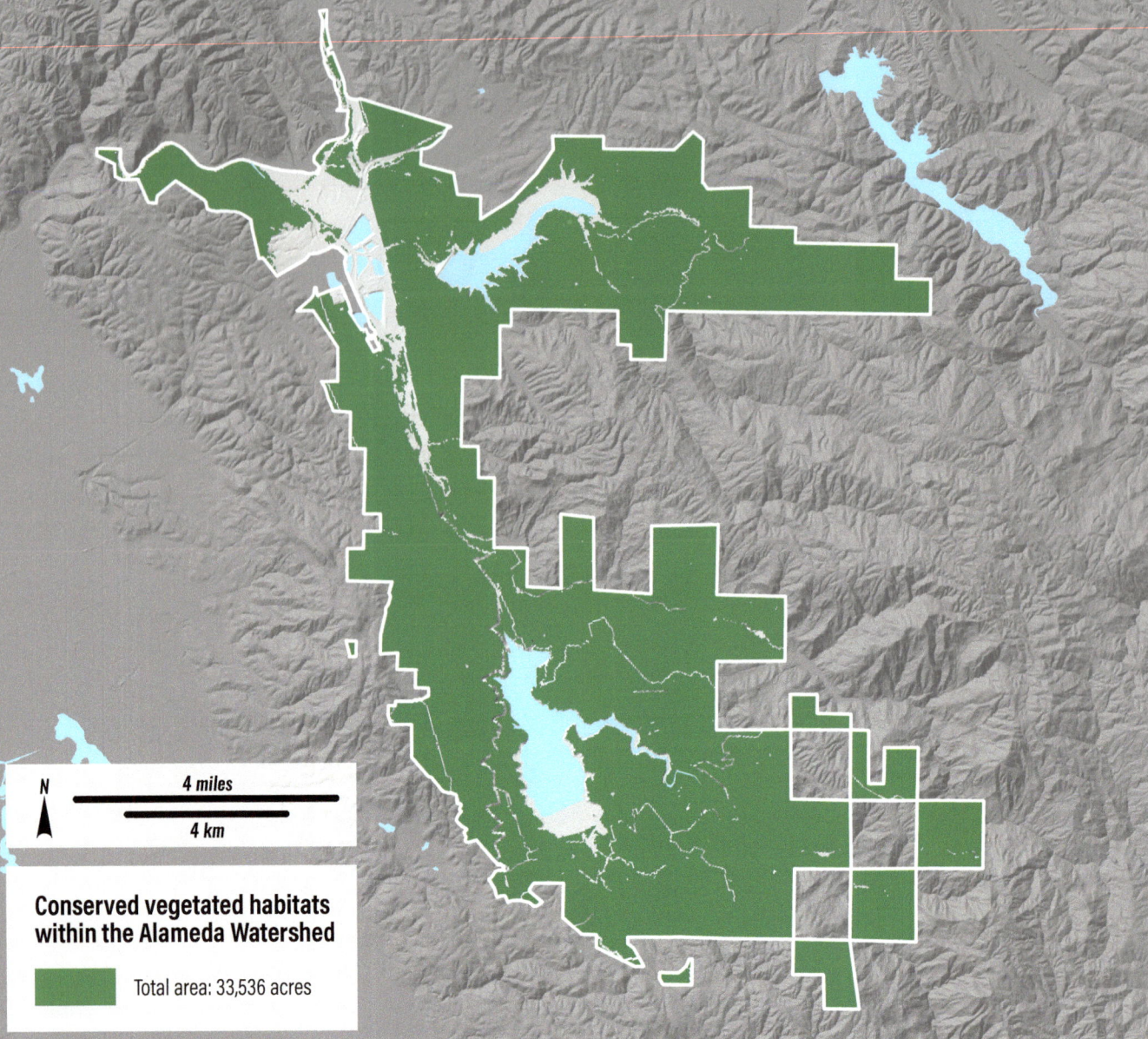

N

4 miles

4 km

**Conserved vegetated habitats within the Alameda Watershed**

Total area: 33,536 acres

**Figure 4.12. Conserved vegetated habitats in the Alameda Watershed.** Map shows all undeveloped land areas within SFPUC Alameda Watershed lands.

**Table 4.1. Estimated avoided carbon losses from urban and agricultural development associated with open space conservation in the Alameda Creek watershed.** For urbanization, avoided carbon losses from biomass were calculated as the difference between average tree and shrub carbon storage in urban sites from Beller et al., 2020 and McHale et al., 2017 and calculated Alameda Watershed carbon stocks (see Chapter 3). Effects of urban development on soil carbon stocks are variable and not well understood (Nowak, 1994). Estimates for conversion to cropland use the IPCC tier 1 default value for cropland biomass carbon (IPCC, 2006) and a range of soil carbon loss rates from Guo and Gifford, 2002 and Murty et al., 2002. Estimates for conversion to vineyard are based on average biomass carbon stocks in mature vineyards from Williams et al., 2020 and soil carbon stocks reported in Williams et al., 2011.

| Ecosystem type | Avoided carbon losses due to land conversion (MT C/acre) | | |
| --- | --- | --- | --- |
| | Urbanization | Conversion to cropland | Conversion to vineyard |
| Watershed-wide average (33,535 acres) | Biomass = 10<br>Soil = unknown | Biomass = 14<br>Soil = 12-24<br>**Total = 26-38** | Biomass = 12<br>Soil = 7-9<br>**Total = 19-21** |
| Grassland (12,744 acres) | Biomass = -5<br>Soil = unknown | Biomass = -1<br>Soil = 9-19<br>**Total = 8-18** | Biomass = -3<br>Soil = 6-7<br>**Total = 3-4** |
| Coastal scrub (1,815 acres) | Biomass = 0<br>Soil = unknown | Biomass = 4<br>Soil = 13-27<br>**Total = 17-31** | Biomass = 2<br>Soil = 8-10<br>**Total = 10-12** |
| Chaparral (4,777 acres) | Biomass = 7<br>Soil = unknown | Biomass = 11<br>Soil = 13-27<br>**Total = 25-38** | Biomass = 10<br>Soil = 8-10<br>**Total = 18-20** |
| Oak savanna (3,988 acres) | Biomass = 15<br>Soil = unknown | Biomass = 19<br>Soil = 13-27<br>**Total = 33-46** | Biomass = 17<br>Soil = 8-10<br>**Total = 25-27** |
| Oak woodland (9,557 acres) | Biomass = 27<br>Soil = unknown | Biomass = 32<br>Soil = 13-27<br>**Total = 45-58** | Biomass = 30<br>Soil = 8-10<br>**Total = 38-40** |
| Riparian forest (653 acres) | Biomass = 58<br>Soil = unknown | Biomass = 62<br>Soil = 13-27<br>**Total = 76-89** | Biomass = 60<br>Soil = 8-10<br>**Total = 68-70** |

# 5 CONCLUSION

Encompassing 39,000 acres of protected open space, the Alameda Watershed maintains substantial carbon stocks in both soils and vegetation. Within the 33,534 acres of grasslands, shrublands, and woodlands evaluated in this study, the watershed stores an estimated 2,500,000 MT C above- and belowground. Per-acre carbon storage varies substantially among ecosystem types, from an estimated $47.2 \pm 0.14$ MT C/acre in grasslands to $128.5 \pm 11.8$ MT C/acre in riparian forests (see Chapter 3).

There are a range of options available to the for managing and enhancing carbon storage in the Alameda Watershed, which are assessed in Chapter 4. Management strategies vary substantially in terms of projected carbon benefits, overall impact on GHG emissions, ecological co-benefits and tradeoffs, potential spatial footprint, and level of certainty and risk, as well as economic cost (which was not assessed in this report). **Native grassland restoration** has the lowest estimated per-acre carbon sequestration benefit of the strategies considered (~0.093 MT C/acre-yr in degraded or unvegetated sites). Three strategies have low to moderate estimated per-acre carbon benefits, including **compost application** (~0.09–0.3 MT C/acre-yr if compost is reapplied every 10 years), **silvopasture** (~0.05–0.35 MT C/acre-yr over 60 years), and **cattle exclusion** (0.08–0.21 MT C/acre-yr in woody vegetation and soil). Strategies with high estimated per-acre carbon sequestration benefits include **riparian restoration** (~0.61–1 MT C/acre-yr averaged over 50 years) and **open space conservation** (~5–90 MT C/acre). The carbon benefits associated with any particular strategy depend heavily on the specific management and maintenance practices employed (e.g., frequency of compost application or density of tree plantings) and the carbon cost of implementing the project, as well as the geographic setting and the time horizon in question.

Our analysis suggests that the carbon management strategies considered here could provide modest advances toward San Francisco's climate action target. Applied to the full potential footprint within the Alameda Watershed, restoring riparian forests could achieve a potential maximum GHG benefit of 5,600–9,100 MT $CO_2$e/yr; planting 6 trees per acre in grazed grasslands (high-density silvopasture) could offset 16,000 MT $CO_2$e/yr; and decadal compost applications could provide a maximum watershed benefit of 480–2,000 MT $CO_2$e/yr on slopes less than 8%, or 1,100–5,000 MT $CO_2$e/yr on slopes up to 20%. If these three management strategies were applied simultaneously, the high-end theoretical GHG benefit of ~30,000 MT $CO_2$e/yr would offset approximately 0.4% of San Francisco's 1990 emissions. However, while this high-end estimate offers a sense of scale, actually maximizing GHG benefits watershed-wide would face a number of feasibility challenges, as outlined in Table 5.1. Additionally, although managing the watershed's ecosystems to promote carbon sequestration can support SFPUC management goals in certain ecological contexts, attempting to maximize carbon gains watershed-wide would likely come at the expense of overall ecosystem health and resilience. Given the complexities, uncertainties, and tradeoffs inherent in decisions around carbon management, a prudent starting point is to identify "low-regrets" management strategies that sequester carbon with significant co-benefits and relatively few risks or potential negative impacts. These strategies will not necessarily *maximize* carbon gains, but will help ensure that carbon management is compatible with other management goals for the watershed, including native biodiversity preservation, fire risk reduction, and protection of water quality and water supply. Table 5.1 provides a generalized overview of the carbon sequestration capacity, co-benefits, and tradeoffs of the strategies presented in Chapter 4, to aid comparison among strategies and identification of low-regrets options. In practice, applying this multi-benefit, low-regrets framework will often be site dependent; management strategies with few tradeoffs in one setting may have significant risks or impacts in other settings.

**Table 5.1. Comparison of key considerations across management strategies discussed in Chapter 4.** For a given consideration, green indicates strong support for the use of a management strategy, red indicates strong concerns, and orange indicates a low to moderate degree of support or concern.

| | Carbon and GHG benefits | Co-benefits | Tradeoffs | Feasibility |
|---|---|---|---|---|
| Rangeland compost | Low to moderate per-acre carbon benefits* | Likely benefits: forage production, soil quality, soil water retention | Key concerns: native biodiversity, residual dry matter control, water quality | Low to moderate concern: access to remote or steep sites |
| Riparian restoration | High per-acre carbon benefits** | Likely benefits: native biodiversity, soil quality, water quality | Moderate concerns: native biodiversity (including risk of pathogen introduction), water supply, wildfire risk, cattle water access | Moderate concern: tree establishment and survival, fencing maintenance, limited opportunity space |
| Silvopasture | Low to moderate per-acre carbon benefits** | Likely benefits: shading, soil quality | Moderate concerns: forage production, native biodiversity (including risk of pathogen introduction), water supply | Moderate concern: tree establishment and survival |
| Cattle exclusion | Low to moderate per-acre carbon benefits** | Potential benefit: water quality | Key concerns: agriculture, wildfire risk, native biodiversity | Moderate concern: fencing maintenance |
| Grassland restoration | Low, uncertain carbon benefits | Likely benefit: native biodiversity | Moderate concern: risk of pathogen introduction if container stock is used | Key concern: likelihood of restoration success |
| Open space consevation | High per-acre carbon benefits | Likely benefits: recreation, agriculture, native biodiversity, water quality, soil quality | Low concern: opportunities for alternate land uses | No major concerns |

**KEY** *Support for the use of a given strategy as a natural climate solution*

| | |
|---|---|
| ⬛ | Low benefits or substantial concerns |
| ⬛ | Moderate benefits or concerns |
| ⬛ | High benefits or low concerns |

*Assumes the material applied is composted green or animal waste. If biosolids are used, assumes material is amended to increase C:N ratios and limit $N_2O$ emissions

**Potential increased wildfire risk may decrease sustainability of carbon benefits.

One management strategy—continued open space conservation—stands out as a low-regrets (or no-regrets) strategy, both in terms of its importance for carbon storage, its numerous co-benefits, and its lack of major downsides. While open space conservation does not represent an additional or novel management activity in the watershed, it is important to recognize the substantial benefits that protection of existing ecosystems and carbon stocks provides. In fact, the benefit of protecting existing carbon stocks exceeds the maximum potential benefits of any other strategies to restore or enhance carbon stocks. Additional land acquisition in the future would only increase the carbon benefit (and other co-benefits) associated with open space conservation in the Alameda Watershed.

Riparian restoration has substantial and well-documented carbon benefits, as well as numerous co-benefits for native biodiversity, soil quality, and water quality. Land use and hydrologic changes over the past 150 years have eliminated large areas of riparian forest in areas such as Sunol Valley and Upper Alameda Creek, and our preliminary assessment suggests that there may be up to ~3,800 acres of suitable area for riparian restoration within the watershed. While there are potential risks and tradeoffs associated with pathogen introduction, decreased water yield and livestock access to water, and increased wildfire risk, the benefits of riparian restoration likely significantly outweigh the risks in many cases and make this a low-regrets management strategy.

Silvopasture, while less effective as a carbon sequestration strategy than riparian restoration on a per-acre basis, also confers clear and predictable carbon benefits and could be a low-regrets strategy if carefully applied in suitable settings. While high density tree plantings would likely increase wildfire risk, sparse plantings emulating historical savanna densities may not substantially alter fire regimes, particularly if sited in areas likely to be more resilient to drought. Silvopasture may also provide important co-benefits to native biodiversity, soil quality, and livestock, with relatively minimal potential tradeoffs. However, the survivorship of tree plantings may be a limiting factor, especially under drought conditions and a changing climate.

Native grassland restoration is another low-regrets management strategy, with no major tradeoffs and significant co-benefits for native biodiversity. The potential carbon benefit associated with native grassland restoration, however, is likely modest and is not well studied outside of degraded or sparsely vegetated sites. In addition, successful native grassland restoration can be difficult to achieve.

Compost application and livestock exclusion, while potentially appropriate and beneficial in certain parts of the watershed, may have more significant tradeoffs that need to be carefully evaluated. Compost application may reduce native grassland biodiversity, promote invasive species (though studies have reported mixed results), and has the potential to impact water quality by increasing nutrient runoff if applied in areas with steep topography. Given these and other tradeoffs and the low estimated per-acre carbon sequestration benefit of compost application relative to other management strategies, we recommend a cautious approach beginning with small-scale pilot studies to investigate the benefits and impacts of this strategy.

A primary objective of livestock grazing in the watershed is fuel management and fire risk reduction, and widespread exclusion of livestock grazing would likely result in substantial woody vegetation encroachment and increased risk of high severity wildfire. In addition, livestock grazing may play an important role in maintaining native grassland biodiversity in some parts of the watershed. However, livestock exclusion may be compatible with both fire management and biodiversity conservation goals in areas that were historically dominated by woody vegetation types or where an increase in woody biomass is desirable and where land use intensity and ignition probability are low. Before excluding livestock from large areas of the watershed, a thorough study of site potential for woody plant encroachment and potential for biodiversity co-benefits or tradeoffs should be conducted.

Because of uncertainties associated with both the potential carbon benefits and the co-benefits and tradeoffs of each management strategy, an adaptive management approach is highly recommended in order to systematically monitor and assess the effects of each strategy in an experimental framework and update management practices based on monitoring results. Pilot studies should be employed to test the effects of management strategies at a small scale before strategies are broadly applied across the watershed. Well designed pilot studies should account for the fact that effects may be highly dependent on the specific site, scale of application, and duration of monitoring. The following bullets highlight several key uncertainties for each of the carbon management strategies that warrant further research, along with ideas for metrics to monitor in pilot projects to indicate how each strategy performs with respect to carbon, co-benefits, and tradeoffs; other management questions or monitoring metrics may be identified in addition to the examples provided:

- **Compost application.** What is the effect of compost application on carbon sequestration in nonnative annual grasslands in the Alameda Watershed? What is the effect of compost application on native grassland biodiversity and water quality? How do repeat compost applications influence long-term carbon sequestration rates? Potential monitoring metrics for pilot projects include GHG fluxes (particularly $CO_2$ exchange and $N_2O$ emissions), forage production, vegetation composition and cover, RDM, changes in cattle stocking rates associated with changes in forage production, soil carbon and nitrogen contents, and runoff nitrate.

- **Riparian restoration.** What is the rate of tree growth? What are the effects of riparian restoration on water yield, water quality and aquatic habitat? Potential monitoring metrics for pilot projects or existing restoration sites include biomass and soil carbon, changes in cattle use of stock ponds or other water sources, and aquatic macroinvertebrates.

- **Silvopasture.** What is the rate of survivorship of planted trees in different settings within the watershed? What is the rate of tree growth? How does tree planting affect soil carbon stocks over time? What planting density optimizes benefits for carbon, livestock, and soils while minimizing tradeoffs? Potential monitoring metrics for pilot projects include tree growth and survivorship, biomass and soil carbon stocks, and forage production.

- **Livestock exclusion.** What is the effect of livestock exclusion on successional transitions between grassland, shrublands, and woodlands in different settings? What long-term rates of carbon sequestration (>25 years) does livestock exclusion achieve? What are the effects of livestock exclusion on native and special-status species occurring in grasslands? Potential monitoring metrics for pilot projects include wildlife presence and abundance (note that some effects on wildlife may not be detectable with small-scale pilot projects), vegetation composition and cover, and biomass and soil carbon stocks.

- **Native grassland restoration.** What is the effect of native grassland restoration on carbon sequestration in different settings (e.g., degraded rangeland, barren ground, nonnative annual grassland)? What is the likelihood of successful restoration outcomes? Potential monitoring metrics for pilot projects include vegetation composition and cover and soil carbon storage.

Research to address these uncertainties may expand the suite of low-regrets carbon management strategies available to SFPUC, or conversely may provide further evidence for potential tradeoffs associated with certain strategies.

## LONG-TERM RESILIENCE

An overarching area of uncertainty is the resilience of carbon gains over the long term, particularly with respect to wildfire and drought, both of which are expected to be exacerbated by climate change (Dass et al., 2018). For example, a substantial increase in woody vegetation cover, while likely to enhance carbon stocks over the short term, may increase the risk of high-severity, stand-replacing wildfire, thus threatening the long-term stability of carbon stocks. Likewise, trees such as valley oak and blue oak are likely to become increasingly vulnerable to drought under climate change (Brown et al., 2018; Browne et al., 2019), and large-scale mortality following an extreme drought event could substantially reduce short term carbon gains. An integrated assessment combining ecophysiological models, dynamic vegetation models, and other sources of information would be useful in assessing the likely resilience of oak woodlands and other vegetation communities in the watershed, as well as the likely effects of carbon management strategies on vegetation succession and wildfire risk under different climate change scenarios. §

Looking down on Alameda Creek, photograph by SFEI.

# 6 REFERENCES

[AB-1279] The California Climate Crisis Act, Cal Assembly Bill 1757, 2021.

[AB-1757] California Global Warming Solutions Act of 2006: Natural and Working Lands, Cal. Assembly Bill 1757, 2021.

Abdalla, M., Hastings, A., Chadwick, D.R., Jones, D.L., Evans, C.D., Jones, M.B., Rees, R.M., Smith, P., 2018. Critical review of the impacts of grazing intensity on soil organic carbon storage and other soil quality indicators in extensively managed grasslands. Agric. Ecosyst. Environ. 253, 62–81. https://doi.org/10.1016/j.agee.2017.10.023

[ACRCD] Alameda County Resource Conservation District, [LD Ford] LD Ford, Rangeland Conservation Science, 2018a. San Francisco Public Utilities Commission Alameda Watershed 2018 Rangeland Monitoring Report - Residual Dry Matter. Prepared for the San Francisco Public Utilities Commission.

[ACRCD] Alameda County Resource Conservation District, [LD Ford] LD Ford, Rangeland Conservation Science, 2018b. San Francisco Public Utilities Commission Alameda Watershed 2018 Rangeland Monitoring Report - Spring Species Composition and Production, Alameda Watershed. Prepared for the San Francisco Public Utilities Commission.

Albaladejo, J., Ortiz, R., Garcia-Franco, N., Navarro, A.R., Almagro, M., Pintado, J.G., Martínez-Mena, M., 2013. Land use and climate change impacts on soil organic carbon stocks in semi-arid Spain. J. Soils Sediments 13, 265–277. https://doi.org/10.1007/s11368-012-0617-7

Anbumozhi, V., Radhakrishnan, J., Yamaji, E., 2005. Impact of riparian buffer zones on water quality and associated management considerations. Ecol. Eng. 24, 517–523. https://doi.org/10.1016/j.ecoleng.2004.01.007

Anderson, K., 2005. Tending the wild: Native American Knowledge and the Management of California's Natural Resources. University of California Press, Berkeley, CA.

Ankenbauer, K.J., Loheide, S.P., 2017. The effects of soil organic matter on soil water retention and plant water use in a meadow of the Sierra Nevada, CA: Soil organic matter affects plant water use. Hydrol. Process. 31, 891–901. https://doi.org/10.1002/hyp.11070

Arnold, C.L., Gibbons, C.J., 1996. Impervious Surface Coverage: The Emergence of a Key Environmental Indicator. J. Am. Plann. Assoc. 62, 243–258. https://doi.org/10.1080/01944369608975688

Artis, S., 2011. Managing California's grassland ecosystems for Athene cunicularia hypugaea. Grasslands Summer 2011.

Azar, C., Lindgren, K., Obersteiner, M., Riahi, K., van Vuuren, D.P., den Elzen, K.M.G.J., Möllersten, K., Larson, E.D., 2010. The feasibility of low CO2 concentration targets and the role of bio-energy with carbon capture and storage (BECCS). Clim. Change 100, 195–202. https://doi.org/10.1007/s10584-010-9832-7

Bardgett, R.D., Manning, P., Morriën, E., De Vries, F.T., 2013. Hierarchical responses of plant-soil interactions to climate change: consequences for the global carbon cycle. J. Ecol. 101, 334–343. https://doi.org/10.1111/1365-2745.12043

Bartolome, J.W., Allen-Diaz, B.H., Barry, S., Ford, L.D., Hammond, M., Hopkinson, P., Ratcliff, F., Spiegal, S., White, M.D., 2014. Grazing for biodiversity in Californian Mediterranean grasslands. Rangelands 36, 36–43. https://doi.org/10.2111/Rangelands-D-14-00024.1

Batjes, N.H., 2016. Harmonized soil property values for broad-scale modeling (WISE30sec) with estimates of global soil carbon stocks. Geoderma 269, 61–68. https://doi.org/10.1016/j.geoderma.2016.01.034

Batllori, E., Ackerly, D.D., Moritz, M.A., 2015. A minimal model of fire-vegetation feedbacks and disturbance stochasticity generates alternative stable states in grassland–shrubland–woodland systems. Environ. Res. Lett. 10, 034018. https://doi.org/10.1088/1748-9326/10/3/034018

Bay Area Open Space Council, 2019. The Conservation Lands Network 2.0 Report. Berkeley, CA.

Beller, E.E., Kelly, M., Larsen, L.G., 2020. From savanna to suburb: effects of 160 years of landscape change on carbon storage in Silicon Valley, California. Landsc. Urban Plan. 195, 103712. https://doi.org/10.1016/j.landurbplan.2019.103712

Bendix, J., Commons, M.G., 2017. Distribution and frequency of wildfire in California riparian ecosystems. Environ. Res. Lett. 12, 075008.

Berhe, A.A., Harden, J.W., Torn, M.S., Kleber, M., Burton, S.D., Harte, J., 2012. Persistence of soil organic matter in eroding versus depositional landform positions. J. Geophys. Res. Biogeosciences 117. https://doi.org/10.1029/2011JG001790

Berhe, A.A., Harte, J., Harden, J.W., Torn, M.S., 2007. The significance of the erosion-induced terrestrial carbon sink. BioScience 57, 337–346.

Bernhardt, E., Swiecki, T.J., 2015. Long-Term Performance of Minimum-Input Oak Restoration Plantings. Gen Tech Rep PSW-GTR-251 Berkeley CA US Dep. Agric. For. Serv. Pac. Southwest Res. Stn. 397-406 251, 397–406.

Boix-Fayos, C., de Vente, J., Albaladejo, J., Martínez-Mena, M., 2009. Soil carbon erosion and stock as affected by land use changes at the catchment scale in Mediterranean ecosystems. Agric. Ecosyst. Environ. 133, 75–85. https://doi.org/10.1016/j.agee.2009.05.013

Bond, T.C., Doherty, S.J., Fahey, D.W., Forster, P.M., Berntsen, T., DeAngelo, B.J., Flanner, M.G., Ghan, S., Kärcher, B., Koch, D., Kinne, S., Kondo, Y., Quinn, P.K., Sarofim, M.C., Schultz, M.G., Schulz, M., Venkataraman, C., Zhang, H., Zhang, S., Bellouin, N., Guttikunda, S.K., Hopke, P.K., Jacobson, M.Z., Kaiser, J.W., Klimont, Z., Lohmann, U., Schwarz, J.P., Shindell, D., Storelvmo, T., Warren, S.G., Zender, C.S., 2013. Bounding the role of black carbon in the climate system: A scientific assessment. J. Geophys. Res. Atmospheres 118, 5380–5552. https://doi.org/10.1002/jgrd.50171

Brauman, K.A., Daily, G.C., Duarte, T.K., Mooney, H.A., 2007. The Nature and value of ecosystem services: an overview highlighting hydrologic services. Annu. Rev. Environ. Resour. 32, 67–98. https://doi.org/10.1146/annurev.energy.32.031306.102758

Brown, B.J., McLaughlin, B.C., Blakey, R.V., Morueta-Holme, N., 2018. Future vulnerability mapping based on response to extreme climate events: Dieback thresholds in an endemic California oak. Divers. Distrib. 24, 1186–1198. https://doi.org/10.1111/ddi.12770

Brown, S., Cotton, M., 2011. Changes in soil properties and carbon content following compost application: results of on-farm sampling. Compost Sci. Util. 19, 87–96. https://doi.org/10.1080/1065657X.2011.10736983

Brown, S., Kruger, C., Subler, S., 2008. Greenhouse gas balance for composting operations. J. Environ. Qual. 37, 1396–1410. https://doi.org/10.2134/jeq2007.0453

Brown, S., Kurtz, K., Bary, A., Cogger, C., 2011. Quantifying benefits associated with land application of organic residuals in Washington State. Environ. Sci. Technol. 45, 7451–7458. https://doi.org/10.1021/es2010418

Brown, S., Shoch, D., Pearson, T., Delaney, M., Franco, G., Surles, T., Therkelsen, R.L., 2004. Methods for Measuring and Monitoring Forestry Carbon Projects in California. Winrock International, for the California Energy Commission, PIER Energy-Related Environmental Research, pp.500-04.

Browne, L., Wright, J.W., Fitz-Gibbon, S., Gugger, P.F., Sork, V.L., 2019. Adaptational lag to temperature in valley oak (Quercus lobata) can be mitigated by genome-informed assisted gene flow. Proc. Natl. Acad. Sci. 116, 25179–25185. https://doi.org/10.1073/pnas.1908771116

Buisson, E., Anderson, S., Holl, K.D., Corcket, E., Hayes, G.F., Peeters, A., Dutoit, T., 2008. Reintroduction of Nassella pulchra to California coastal grasslands: effects of topsoil removal, plant neighbour removal and grazing. Appl. Veg. Sci. 11, 195–204.

Bulger, J.B., Scott Jr, N.J., Seymour, R.B., 2003. Terrestrial activity and conservation of adult California red-legged frogs Rana aurora draytonii in coastal forests and grasslands. Biol. Conserv. 110, 85–95.

Bullard, V., Smither-Kopperl, M., 2020. Compost Application Effects on Rangeland Species Composition and Forage Production. Lockeford, California. U.S. Department of Agriculture, Lockeford Plant Materials Center.

Burrell, T.K., O'Brien, J.M., Graham, S.E., Simon, K.S., Harding, J.S., McIntosh, A.R., 2014. Riparian shading mitigates stream eutrophication in agricultural catchments. Freshw. Sci. 33, 73–84. https://doi.org/10.1086/674180

Callaway, R.M., Davis, F.W., 1993. Vegetation dynamics, fire, and the physical environment in coastal central California. Ecology 74, 1567–1578. https://doi.org/10.2307/1940084

Callaway, R.M., Nadkarni, N.M., Mahall, B.E., 1991. Facilitation and interference of Quercus douglasii on understory productivity in central California. Ecology 72, 1484–1499. https://doi.org/10.2307/1941122

Cameron, D.R., Marvin, D.C., Remucal, J.M., Passero, M.C., 2017. Ecosystem management and land conservation can substantially contribute to California's climate mitigation goals. Proc. Natl. Acad. Sci. 114, 12833–12838. https://doi.org/10.1073/pnas.1707811114

Camping, 2002. Proceedings of the Fifth Symposium on Oak Woodlands: Oaks in California's Changing Landscape, October 22-25, 2001, San Diego, California. U.S. Department of Agriculture, Forest Service, Pacific Southwest Research Station.

Canadell, J.G., Raupach, M.R., 2008. Managing forests for climate change mitigation. Science 320, 1456–1457. https://doi.org/10.1126/science.1155458

Canadell, J.G., Schulze, E.D., 2014. Global potential of biospheric carbon management for climate mitigation. Nat. Commun. 5, 5282. https://doi.org/10.1038/ncomms6282

[CARB] California Air Resources Board, 2022a. 2022 Scoping Plan for Achieving Carbon Neutrality.

[CARB] California Air Resources Board, 2022b. Wildfire Emission Estimates for 2021.

Carey, C.J., Gravuer, K., Gennet, S., Osleger, D., Wood, S.A., 2020. Supporting evidence varies for rangeland management practices that seek to improve soil properties and forage production in California. Calif. Agric. 74, 101–111. https://doi.org/10.3733/ca.2020a0015

Cayan, D., Tyree, M., Lacobellis, S., 2012. Climate change scenarios for the San Francisco region. Public Interest Energy Research White Paper. Prepared for the California Energy Commission by Scripps Institution of Oceanography, University of California San Diego.

Cayan, D.R., Maurer, E.P., Dettinger, M.D., Tyree, M., Hayhoe, K., 2008. Climate change scenarios for the California region. Clim. Change 87, 21–42.

Cellier, A., Gauquelin, T., Baldy, V., Ballini, C., 2014. Effect of organic amendment on soil fertility and plant nutrients in a post-fire Mediterranean ecosystem. Plant Soil 376, 211–228. https://doi.org/10.1007/s11104-013-1969-5

Central Coast Wetlands Group, 2017. Development of New Tools to Assess Riparian Extent and Condition - A Central Coast Pilot Study. Final report for USEPA Wetlands Program Development Grant CD- 00T83101.

Chapin III, F.S., McFarland, J., David McGuire, A., Euskirchen, E.S., Ruess, R.W., Kielland, K., 2009. The changing global carbon cycle: linking plant–soil carbon dynamics to global consequences. J. Ecol. 97, 840–850. https://doi.org/10.1111/j.1365-2745.2009.01529.x

Charles, A., Rochette, P., Whalen, J.K., Angers, D.A., Chantigny, M.H., Bertrand, N., 2017. Global nitrous oxide emission factors from agricultural soils after addition of organic amendments: A meta-analysis. Agric. Ecosyst. Environ. 236, 88–98. https://doi.org/10.1016/j.agee.2016.11.021

Chojnacky, D.C., Heath, L.S., Jenkins, J.C., 2014. Updated generalized biomass equations for North American tree species. For. Int. J. For. Res. 87, 129–151. https://doi.org/10.1093/forestry/cpt053

Chou, W.W., Silver, W.L., Jackson, R.D., Thompson, A.W., Allen⊠Diaz, B., 2008. The sensitivity of annual grassland carbon cycling to the quantity and timing of rainfall. Glob. Change Biol. 14, 1382–1394. https://doi.org/10.1111/j.1365-2486.2008.01572.x

Churkina, G., Running, S.W., 1998. Contrasting climatic controls on the estimated productivity of global terrestrial biomes. Ecosystems 1, 206–215.

Ciais, P., Sabine, C., Bala, G., Bopp, L., Brovkin, V., Canadell, J., Chhabra, A., DeFries, R., Galloway, J., Heimann, M., Jones, C., Le Quere, C., Myneni, R.B., Piao, S., Thornton, P., 2013. Carbon and Other Biogeochemical Cycles, in: Climate Change 2013: The Physical Science Basis. Contribution of Working Group I to the Fifth Assessment Report of the Intergovernmental Panel on Climate Change [Stocker, T.F., D. Qin, G.-K. Plattner, M. Tignor, S.K. Allen, J. Boschung, A. Nauels, Y. Xia, V. Bex and P.M Midgley (Eds)]. Cambridge University Press, Cambridge, United Kingdom and New York, NY, USA.

City and County of San Francisco, 2021. San Francisco's Climate Action Plan 2021.

Cleland, E.E., Funk, J.L., Allen, E.B., 2016. Twenty-two. Coastal sage scrub, in: Coastal Sage Scrub. University of California Press, pp. 429–448. https://doi.org/10.1525/9780520962170-026

Clinton, N.E., Gong, P., Scott, K., 2006. Quantification of pollutants emitted from very large wildland fires in Southern California, USA. Atmos. Environ. 40, 3686–3695. https://doi.org/10.1016/j.atmosenv.2006.02.016

Conant, R.T., Cerri, C.E.P., Osborne, B.B., Paustian, K., 2017. Grassland management impacts on soil carbon stocks: a new synthesis. Ecol. Appl. 27, 662–668. https://doi.org/10.1002/eap.1473

Corbin, J.D., D'Antonio, C.M., Bainbridge, S., 2004. Tipping the balance in the restoration of native plants: experimental approaches to changing the exotic: native ratio in California grassland, in: Experimental Approaches to Conservation Biology. University of California Press Berkeley, CA, pp. 154–179.

Cusack, D.F., Kazanski, C.E., Hedgpeth, A., Chow, K., Cordeiro, A.L., Karpman, J., Ryals, R., 2021. Reducing climate impacts of beef production: A synthesis of life cycle assessments across management systems and global regions. Glob. Change Biol. 27, 1721–1736.

Dahlgren, R., Singer, M.J., 1991. Nutrient cycling in managed and unmanaged oak woodland-grass ecosystems. In Symposium on Oak Woodlands and Hardwood Rangeland Management. Gen. Tech. Rep. PSW-GTR-126. Berkeley, CA: Pacific Southwest Research Station, Forest Service, US Department of Agriculture (pp. 337-341).

Dahlgren, R.A., Singer, M.J., Huang, X., 1997. Oak tree and grazing impacts on soil properties and nutrients in a California oak woodland. Biogeochemistry 39, 45–64. https://doi.org/10.1023/A:1005812621312

D'Antonio, C.M., Malmstrom, C., Reynolds, S.A., Gerlach, J., 2007. Species in California Grassland. In California Grasslands: Ecology and Management, Mark Stromberg, Jeffrey Corbin, Carla D'Antonio, eds. University of California Press, Berkeley, CA.

Daryanto, S., Eldridge, D.J., Throop, H.L., 2013. Managing semi-arid woodlands for carbon storage: Grazing and shrub effects on above- and belowground carbon. Agric. Ecosyst. Environ. 169, 1–11. https://doi.org/10.1016/j.agee.2013.02.001

Dass, P., Houlton, B.Z., Wang, Y., Warlind, D., 2018. Grasslands may be more reliable carbon sinks than forests in California. Environ. Res. Lett. 13, 074027. https://doi.org/10.1088/1748-9326/aacb39

Davidson, E.A., Janssens, I.A., 2006. Temperature sensitivity of soil carbon decomposition and feedbacks to climate change. Nature 440, 165–173. https://doi.org/10.1038/nature04514

Davies, K.W., Boyd, C.S., Bates, J.D., Hulet, A., 2015. Winter grazing can reduce wildfire size, intensity and behaviour in a shrub-grassland. Int. J. Wildland Fire 25, 191–199.

Davis, F., Borchert, M., 2006. Central Coast bioregion, in: Fire in California's ecosystems (Eds. N. G. Sugihara, J. W. van Wagtendonk, K. Shaffer, J. Fites-Kaufman, and A. E. Thode. pp. 321–329.

Davis, M.A., Grime, J.P., Thompson, K., 2000. Fluctuating resources in plant communities: a general theory of invasibility. J. Ecol. 88, 528–534. https://doi.org/10.1046/j.1365-2745.2000.00473.x

De Deyn, G.B., Cornelissen, J.H.C., Bardgett, R.D., 2008. Plant functional traits and soil carbon sequestration in contrasting biomes. Ecol. Lett. 11, 516–531. https://doi.org/10.1111/j.1461-0248.2008.01164.x

De Gryze, S., Wolf, A., Kaffka, S.R., Mitchell, J., Rolston, D.E., Temple, S.R., Lee, J., Six, J., 2010. Simulating greenhouse gas budgets of four California cropping systems under conventional and alternative management. Ecol. Appl. 20, 1805–1819.

del Mar Montiel-Rozas, M., Panettieri, M., Madejón, P., Madejón, E., 2016. Carbon sequestration in restored soils by applying organic amendments. Land Degrad. Dev. 27, 620–629.

DeLonge, M.S., Ryals, R., Silver, W.L., 2013. A lifecycle model to evaluate carbon sequestration potential and greenhouse gas dynamics of managed grasslands. Ecosystems 16, 962–979. https://doi.org/10.1007/s10021-013-9660-5

DiTomaso, J.M., 2000. Invasive weeds in rangelands: Species, impacts, and management. Weed Sci. 48, 255–265. https://doi.org/10.1614/0043-1745(2000)048[0255:IWIRSI]2.0.CO;2

Di Vittorio, A.V., Simmonds, M.B., Nico, P., 2021. Quantifying the effects of multiple land management practices, land cover change, and wildfire on the California landscape carbon budget with an empirical model. PLOS ONE 16, e0251346. https://doi.org/10.1371/journal.pone.0251346

Diacono, M., Montemurro, F., 2011. Long-term effects of organic amendments on soil fertility, in: Lichtfouse, E., Hamelin, M., Navarrete, M., Debaeke, P. (Eds.), Sustainable Agriculture Volume 2. Springer Netherlands, Dordrecht, pp. 761–786. https://doi.org/10.1007/978-94-007-0394-0_34

Dosskey, M.G., Vidon, P., Gurwick, N.P., Allan, C.J., Duval, T.P., Lowrance, R., 2010. The role of riparian vegetation in protecting and improving chemical water quality in streams. JAWRA J. Am. Water Resour. Assoc. 46, 261–277. https://doi.org/10.1111/j.1752-1688.2010.00419.x

DuPont, S.T., Culman, S.W., Ferris, H., Buckley, D.H., Glover, J.D., 2010. No-tillage conversion of harvested perennial grassland to annual cropland reduces root biomass, decreases active carbon stocks, and impacts soil biota. Agric. Ecosyst. Environ., Special section Harvested perennial grasslands: Ecological models for farming's perennial future 137, 25–32. https://doi.org/10.1016/j.agee.2009.12.021

Dybala, K.E., Matzek, V., Gardali, T., Seavy, N.E., 2019. Carbon sequestration in riparian forests: A global synthesis and meta-analysis. Glob. Change Biol. 25, 57–67. https://doi.org/10.1111/gcb.14475

[EPA] United States Environmental Protection Agency., 2009. Potential for Reducing Greenhouse Gas Emissions in the Construction Sector. United States Environmental Protection Agency, Washington, DC.

[EPA] United States Environmental Protection Agency., 2022. Greenhouse Gas Equivalencies Calculator, updated March 2022. Available from: https://www.epa.gov/energy/greenhouse-gas-equivalencies-calculator

Eve, M., Pape, D., Flugge, M., Steele, R., Man, D., Riley, M., Biggar, S., 2014. Quantifying greenhouse gas fluxes in agriculture and forestry:  methods for entity-scale inventory (Technical Bulletin Number 1939). US Department of Agriculture, Office of the Chief Economist.

Evett, R.R., Bartolome, J.W., 2013. Phytolith evidence for the extent and nature of prehistoric Californian grasslands. The Holocene 23, 1644–1649.

Eviner, V., Malmstrom, C., 2018. California's native perennial grasses provide strong suppression of goatgrass and medusahead. Grasslands 28, 3–6.

Eviner, V.T., 2016. Grasslands. In: Ecosystems of California. University of California Press, pp. 449–478.

Fehmi, J.S., Russo, S.E., Bartolome, J.W., 2005. The effects of livestock on California ground squirrels (Spermophilus beecheyii). Rangel. Ecol. Manag. 58, 352–359. https://doi.org/10.2111/1551-5028(2005)058[0352:TEOLOC]2.0.CO;2

Follett, R.F., Reed, D.A., 2010. Soil carbon sequestration in grazing lands: societal benefits and policy implications. Rangel. Ecol. Manag. 63, 4–15. https://doi.org/10.2111/08-225.1

Ford L.D., Hayes, G.F., 2007. Northern Coastal Scrub and Coastal Prairie. In Terrestrial Vegetation of California, eds. Michael Barbour, Todd Keeler-Wolf, Allan A. Schoenherr. Berkeley: University of California Press.

Frankel, S.J., Conforti, C., Hillman, J., Ingolia, M., Shor, A., Benner, D., Alexander, J.M., Bernhardt, E., Swiecki, T.J., 2020. Phytophthora introductions in restoration areas: Responding to protect California native flora from human-assisted pathogen spread. Forests 11, 1291.

Galik, C.S., Jackson, R.B., 2009. Risks to forest carbon offset projects in a changing climate. For. Ecol. Manag. 257, 2209–2216.

Gasser, T., Guivarch, C., Tachiiri, K., Jones, C.D., Ciais, P., 2015. Negative emissions physically needed to keep global warming below 2 °C. Nat. Commun. 6, 7958. https://doi.org/10.1038/ncomms8958

Gea-Izquierdo, G., Gennet, S., Bartolome, J.W., 2007. Assessing plant-nutrient relationships in highly invaded Californian grasslands using non-normal probability distributions. Appl. Veg. Sci. 10, 343–350.

Gennet, S., Spotswood, E., Hammond, M., Bartolome, J.W., 2017. Livestock grazing supports native plants and songbirds in a California annual grassland. PLoS ONE 12. https://doi.org/10.1371/journal.pone.0176367

Gibbs, H.K., Ruesch, A., 2008. New IPCC tier-1 Global Biomass Carbon Map for the Year 2000. https://doi.org/10.15485/1463800

Goals Project, 1999. Baylands Ecosystem Habitat Goals. A report of habitat recommendations prepared by the San Francisco Bay Area Wetlands Ecosystem Goals Project. U.S. Environmental Protection Agency, San Francisco, Calif./S.F. Bay Regional Water Quality Control Board, Oakland, Calif.

Golet, G.H., Gardali, T., Howell, C.A., Hunt, J., Luster, R.A., Rainey, W., Roberts, M.D., Silveira, J., Swagerty, H., Williams, N., 2008. Wildlife response to riparian restoration on the Sacramento River. San Franc. Estuary Watershed Sci. 6.

Golubiewski, N.E., 2006. Urbanization increases grassland carbon pools: effects of landscaping in Colorado's Front Range. Ecol. Appl. 16, 555–571. https://doi.org/10.1890/1051-0761(2006)016[0555:UIGCPE]2.0.CO;2

Gonzalez, P., Asner, G.P., Battles, J.J., Lefsky, M.A., Waring, K.M., Palace, M., 2010. Forest carbon densities and uncertainties from Lidar, QuickBird, and field measurements in California. Remote Sens. Environ. 114, 1561–1575. https://doi.org/10.1016/j.rse.2010.02.011

Gonzalez, P., Battles, J.J., Collins, B.M., Robards, T., Saah, D.S., 2015. Aboveground live carbon stock changes of California wildland ecosystems, 2001–2010. For. Ecol. Manag. 348, 68–77. https://doi.org/10.1016/j.foreco.2015.03.040

Gornish, E.S., Ambrozio dos Santos, P., 2016. Invasive species cover, soil type, and grazing interact to predict long-term grassland restoration success. Restor. Ecol. 24, 222–229.

Goss, M., Swain, D.L., Abatzoglou, J.T., Sarhadi, A., Kolden, C.A., Williams, A.P., Diffenbaugh, N.S., 2020. Climate change is increasing the likelihood of extreme autumn wildfire conditions across California. Environ. Res. Lett. 15, 094016. https://doi.org/10.1088/1748-9326/ab83a7

Gravuer, K., Gennet, S., Throop, H.L., 2019. Organic amendment additions to rangelands: A meta☒analysis of multiple ecosystem outcomes. Glob. Change Biol. 25, 1152–1170. https://doi.org/10.1111/gcb.14535

Gravuer, K., Gunasekara, A., 2016. Compost Application Rates for California Croplands and Rangelands for a CDFA Healthy Soils Incentives Program. Sacramento, California.

Gregorich, E.G., Greer, K.J., Anderson, D.W., Liang, B.C., 1998. Carbon distribution and losses: erosion and deposition effects. Soil Tillage Res. 47, 291–302. https://doi.org/10.1016/S0167-1987(98)00117-2

Grossinger, R.M., Beller, E.E., Salomon, M.N., Whipple, A.A., Askevold, R.A., Striplen, C.J., Brewster, E., Leidy, R.A., 2008. South Santa Clara Valley Historical Ecology Study, including Soap Lake, the Upper Pajaro River, and Llagas, Uvas-Carnadero, and Pacheco Creeks Leidy (Historical Ecology Program Report No. 558). San Francisco Estuary Institute, Oakland, CA.

Guo, L.B., Gifford, R.M., 2002. Soil carbon stocks and land use change: a meta analysis. Glob. Change Biol. 8, 345–360. https://doi.org/10.1046/j.1354-1013.2002.00486.x

Gyssels, G., Poesen, J., Bochet, E., Li, Y., 2005. Impact of plant roots on the resistance of soils to erosion by water: a review. Prog. Phys. Geogr. Earth Environ. 29, 189–217. https://doi.org/10.1191/0309133305pp443ra

Hanson, W.D., Kukula, F., Nelson, C., Williams, M., 2005. San Francisco Water and Power : a History of the Municipal Water Department & Hetch Hetchy System. [San Francisco, Calif. : Public Utilities Commission, City and County of San Francisco].

Hargreaves, J., Adl, M., Warman, P., 2008. A review of the use of composted municipal solid waste in agriculture. Agric. Ecosyst. Environ. 123, 1–14. https://doi.org/10.1016/j.agee.2007.07.004

Harpole, W.S., Sullivan, L.L., Lind, E.M., Firn, J., Adler, P.B., Borer, E.T., Chase, J., Fay, P.A., Hautier, Y., Hillebrand, H., 2016. Addition of multiple limiting resources reduces grassland diversity. Nature 537, 93–96.

Harrison SP, Viers, J.H., 2007. Serpentine Grasslands. In California Grasslands: Ecology and Management, eds. Mark Stromberg, Jeffrey Corbin, Carla D'Antonio. Berkeley, Calif: University of California Press.

Haug, E.A., Oliphant, L.W., 1990. Movements, activity patterns, and habitat use of burrowing owls in Saskatchewan. J. Wildl. Manag. 54, 27–35. https://doi.org/10.2307/3808896

Hayes, G.F., Holl, K.D., 2003. Cattle grazing impacts on annual forbs and vegetation composition of mesic grasslands in California. Conserv. Biol. 17, 1694–1702. https://doi.org/10.1111/j.1523-1739.2003.00281.x

Herbst, D.B., Knapp, R.A., 1995. Evaluation of Rangeland Stream Condition and Recovery using Physical and Biological Assessments of Nonpoint Source Pollution. Technical completion report, project number UCAL-WRC-W-818

Herman, D.J., Halverson, L.J., Firestone, M.K., 2003. Nitrogen dynamics in an annual grassland: oak canopy, climate, and microbial population effects. Ecol. Appl. 13, 593–604. https://doi.org/10.1890/1051-0761(2003)013[0593:NDIAAG]2.0.CO;2

Hernando, S., Lobo, M.C., Polo, A., 1989. Effect of the application of a municipal refuse compost on the physical and chemical properties of a soil. Sci. Total Environ. 81–82, 589–596. https://doi.org/10.1016/0048-9697(89)90167-8

Hilty, J.A., Merenlender, A.M., 2004. Use of riparian corridors and vineyards by mammalian predators in northern California. Conserv. Biol. 18, 126–135. https://doi.org/10.1111/j.1523-1739.2004.00225.x

Hobbs, R.J., Mooney, H.A., 1986. Community changes following shrub invasion of grassland. Oecologia 70, 508–513. https://doi.org/10.1007/BF00379896

Hope, A., Tague, C., Clark, R., 2007. Characterizing post‐fire vegetation recovery of California chaparral using TM/ETM+ time‐series data. Int. J. Remote Sens. 28, 1339–1354. https://doi.org/10.1080/01431160600908924

Horner, G.J., Baker, P.J., Nally, R.M., Cunningham, S.C., Thomson, J.R., Hamilton, F., 2010. Forest structure, habitat and carbon benefits from thinning floodplain forests: managing early stand density makes a difference. For. Ecol. Manag. 259, 286–293. https://doi.org/10.1016/j.foreco.2009.10.015

Huntsinger, L., Bartolome, J.W., 1992. Ecological dynamics of Quercus dominated woodlands in California and southern Spain: a state-transition model. Vegetatio 99, 299–305.

Huntsinger, L., Bartolome, J.W., D'Antonio, C.M., 2007. Grazing management on California's Mediterranean grasslands, in: California Grasslands: Ecology and Management. University of California Press. https://doi.org/10.1525/california/9780520252202.001.0001

Huston, M.A., Marland, G., 2003. Carbon management and biodiversity. J. Environ. Manage., Maintaining Forest Biodiversity 67, 77–86. https://doi.org/10.1016/S0301-4797(02)00190-1

Hutyra, L., Yoon, B., Alberti, M., 2011. Terrestrial carbon stocks across a gradient of urbanization: a study of the Seattle, WA region. Glob. Change Biol. 17, 783–797. https://doi.org/10.1111/j.1365-2486.2010.02238.x

IPCC, 2006. Guidelines for National Greenhouse Gas Inventories, prepared by the National Greenhouse Gas Inventories Programme, Eggleston H.S., Buendia L., Miwa K., Ngara T., and Tanabe K. (eds). IGES, Japan.

Jacinthe, P.A., 2015. Carbon dioxide and methane fluxes in variably-flooded riparian forests. Geoderma 241–242, 41–50. https://doi.org/10.1016/j.geoderma.2014.10.013

Jacinthe, P.A., Vidon, P., Fisher, K., Liu, X., Baker, M.E., 2015. Soil Methane and Carbon Dioxide Fluxes from Cropland and Riparian Buffers in Different Hydrogeomorphic Settings. J. Environ. Qual. 44, 1080–1090. https://doi.org/10.2134/jeq2015.01.0014

Jackson, L.E., Potthoff, M., Steenwerth, K.L., O'Geen, A.T., Stromberg, M.R., Scow, K.M., 2007. Soil biology and carbon sequestration in grasslands. In Ecology and Management of California Grasslands. Univ. Calif. Press Berkeley CA 107–118.

Jackson, R.B., Jobbágy, E.G., Avissar, R., Roy, S.B., Barrett, D.J., Cook, C.W., Farley, K.A., le Maitre, D.C., McCarl, B.A., Murray, B.C., 2005. Trading water for carbon with biological carbon sequestration. Science 310, 1944–1947. https://doi.org/10.1126/science.1119282

Jones & Stokes Associates, Inc. 2003. [Vegetation mapping SFPUC Alameda Watershed Habitat Conservation Plan -Biological Inventory Report.] Data provided by San Francisco Public Utilities Commission

Jones and Stokes, 2008. East Bay Regional Parks District Carbon Sequestration Evaluation. Prepared for the East Bay Regional Parks District.

Jose, S., 2009. Agroforestry for ecosystem services and environmental benefits: an overview. Agrofor. Syst. 76, 1–10.

Keeley, J.E., 2005. Fire history of the San Francisco East Bay region and implications for landscape patterns. Int. J. Wildland Fire 14, 285. https://doi.org/10.1071/WF05003

Keeley, J.E., Bond, W.J., Bradstock, R.A., Pausas, J.G., Rundel, P.W., 2011. Fire in California, in: Fire in Mediterranean Ecosystems: Ecology, Evolution and Management. Cambridge University Press, Cambridge. https://doi.org/10.1017/CBO9781139033091

Keeley, J.E., Fotheringham, C.J., Baer-Keeley, M., 2005. Determinants of postfire recovery and succession in Mediterranean-climate shrublands of California. Ecol. Appl. 15, 1515–1534. https://doi.org/10.1890/04-1005

Keeley, J.E., Syphard, A., 2016. Climate change and future fire regimes: examples from California. Geosciences 6, 37. https://doi.org/10.3390/geosciences6030037

Kinoshita, A.M., Hogue, T.S., 2011. Spatial and temporal controls on post-fire hydrologic recovery in Southern California watersheds. Catena 87, 240–252.

Kobziar, L.N., McBride, J.R., 2006. Wildfire burn patterns and riparian vegetation response along two northern Sierra Nevada streams. For. Ecol. Manag. 222, 254–265.

Koteen, L.E., Baldocchi, D.D., Harte, J., 2011. Invasion of non-native grasses causes a drop in soil carbon storage in California grasslands. Environ. Res. Lett. 6, 044001. https://doi.org/10.1088/1748-9326/6/4/044001

Kueppers, L.M., Snyder, M.A., Sloan, L.C., Zavaleta, E.S., Fulfrost, B., 2005. Modeled regional climate change and California endemic oak ranges. Proc. Natl. Acad. Sci. 102, 16281–16286. https://doi.org/10.1073/pnas.0501427102

Lal, R., 2020. Soil organic matter and water retention. Agron. J. 112, 3265–3277. https://doi.org/10.1002/agj2.20282

Lal, R., 2016. Soil health and carbon management. Food Energy Secur. 5, 212–222. https://doi.org/10.1002/fes3.96

Lenihan, J.M., Bachelet, D., Neilson, R.P., Drapek, R., 2008. Response of vegetation distribution, ecosystem productivity, and fire to climate change scenarios for California. Clim. Change 87, 215–230.

Lennox, M.S., Lewis, D.J., Jackson, R.D., Harper, J., Larson, S., Tate, K.W., 2011. Development of vegetation and aquatic habitat in restored riparian sites of California's north coast rangelands. Restor. Ecol. 19, 225–233. https://doi.org/10.1111/j.1526-100X.2009.00558.x

Lewis, D.J., Lennox, M., O'Green, A., Creque, J., Eviner, V., Larson, S., Harper, J., Doran, M., Tate, K.W., 2015. Creek Carbon: Mitigating Greenhouse Gas Emissions Through Riparian Revegetation. University of California Cooperative Extension in Marin County, Novato, California.

Luong, J.C., Turner, P.L., Phillipson, C.N., Seltmann, K.C., 2019. Local grassland restoration affects insect communities. Ecol. Entomol. 44, 471–479. https://doi.org/10.1111/een.12721

Lydersen, J.M., Collins, B.M., Miller, J.D., Fry, D.L., Stephens, S.L., 2016. Relating fire-caused change in forest structure to remotely sensed estimates of fire severity. Fire Ecol. 12, 99–116. https://doi.org/10.4996/fireecology.1203099

Ma, S., Baldocchi, D.D., Xu, L., Hehn, T., 2007. Inter-annual variability in carbon dioxide exchange of an oak/grass savanna and open grassland in California. Agric. For. Meteorol. 147, 157–171.

Mackay, J.E., Cunningham, S.C., Cavagnaro, T.R., 2016. Riparian reforestation: are there changes in soil carbon and soil microbial communities? Sci. Total Environ. 566–567, 960–967. https://doi.org/10.1016/j.scitotenv.2016.05.045

Malhi, Y., Wood, D., Baker, T.R., Wright, J., Phillips, O.L., Cochrane, T., Meir, P., Chave, J., Almeida, S., Arroyo, L., Higuchi, N., Killeen, T.J., Laurance, S.G., Laurance, W.F., Lewis, S.L., Monteagudo, A., Neill, D.A., Vargas, P.N., Pitman, N.C.A., Quesada, C.A., Salomão, R., Silva, J.N.M., Lezama, A.T., Terborgh, J., Martínez, R.V., Vinceti, B., 2006. The regional variation of aboveground live biomass in old-growth Amazonian forests: biomass in Amazonian forests. Glob. Change Biol. 12, 1107–1138. https://doi.org/10.1111/j.1365-2486.2006.01120.x

Malmstrom, C.M., Butterfield, H.S., Planck, L., Long, C.W., Eviner, V.T., 2017. Novel fine-scale aerial mapping approach quantifies grassland weed cover dynamics and response to management. PLOS ONE 12, e0181665. https://doi.org/10.1371/journal.pone.0181665

Marañón, T., Bartolome, J.W., 1994. Coast live oak (Quercus agrifolia) effects on grassland biomass and diversity. Madroño 41, 39–52.

Matzek, V., Lewis, D., O'Geen, A., Lennox, M., Hogan, S.D., Feirer, S.T., Eviner, V., Tate, K.W., 2020. Increases in soil and woody biomass carbon stocks as a result of rangeland riparian restoration. Carbon Balance Manag. 15, 16. https://doi.org/10.1186/s13021-020-00150-7

Matzek, V., Stella, J., Ropion, P., 2018. Development of a carbon calculator tool for riparian forest restoration. Appl. Veg. Sci. 21, 584–594. https://doi.org/10.1111/avsc.12400

Matzek, V., Warren, S., Fisher, C., 2016. Incomplete recovery of ecosystem processes after two decades of riparian forest restoration. Restor. Ecol. 24, 637–645. https://doi.org/10.1111/rec.12361

Mayer, A., Silver, W.L., 2022. The climate change mitigation potential of annual grasslands under future climates. Ecol. Appl. e2705.

Mayer, P.M., Reynolds, S.K., Canfield, T.J., 2005. Riparian Buffer Width, Vegetative Cover, and Nitrogen Removal Effectiveness: A Review of Current Science and Regulations. United States Environmental Protection Agency office of Research and Development.

McBride, J., Heady, H.F., 1968. Invasion of grassland by Baccharis pilularis DC. J. Range Manag. 21, 106. https://doi.org/10.2307/3896366

McBride, J.R., 1974. Plant succession in the Berkeley hills, California. Madroño 22, 317–329.

McHale, M., Hall, S., Majumdar, A., Grimm, N., 2017. Carbon lost and carbon gained: A study of vegetation and carbon tradeoffs among diverse land uses in the Phoenix, AZ. Ecol. Appl. Publ. Ecol. Soc. Am. 27. https://doi.org/10.1002/eap.1472

McMichael, C.E., Hope, A.S., Roberts, D.A., Anaya, M.R., 2004. Post-fire recovery of leaf area index in California chaparral: A remote sensing-chronosequence approach. Int. J. Remote Sens. 25, 4743–4760. https://doi.org/10.1080/01431160410001726067

McSherry, M.E., Ritchie, M.E., 2013. Effects of grazing on grassland soil carbon: a global review. Glob. Change Biol. 19, 1347–1357. https://doi.org/10.1111/gcb.12144

Mead, A., Peñaloza Ramirez, J., Bartlett, M.K., Wright, J.W., Sack, L., Sork, V.L., 2019. Seedling response to water stress in valley oak (Quercus lobata) is shaped by different gene networks across populations. Mol. Ecol. 28, 5248–5264. https://doi.org/10.1111/mec.15289

Mensing, S.A., 1998. 560 years of vegetation change in the region of Santa Barbara, California. Madroño 45, 1–11.

Miller, J.D., Knapp, E.E., Key, C.H., Skinner, C.N., Isbell, C.J., Creasy, R.M., Sherlock, J.W., 2009. Calibration and validation of the relative differenced Normalized Burn Ratio (RdNBR) to three measures of fire severity in the Sierra Nevada and Klamath Mountains, California, USA. Remote Sens. Environ. 113, 645–656. https://doi.org/10.1016/j.rse.2008.11.009

Miller, J.D., Thode, A.E., 2007. Quantifying burn severity in a heterogeneous landscape with a relative version of the delta Normalized Burn Ratio (dNBR). Remote Sens. Environ. 109, 66–80. https://doi.org/10.1016/j.rse.2006.12.006

Milliken, R., 1995. A Time of Little Choice: The Disintegration of Tribal Culture in the San Francisco Bay Area, 1769-1810. Ballena Press.

Milliken, R., Johnson, John, Earle, David, Smith, Norval, Mikkelsen, Patricia, Brandy, Paul, King, Jerome, 2010. Contact-Period Native California Community Distribution Model [WWW Document]. URL http://wiki.farwestern.com/index.php?title=VOLUME_1 (accessed 10.26.22).

Minnich, R.A., 2008. California's Fading Wildflowers: Lost Legacy and Biological Invasions. Univ of California Press.

Minx, J.C., Lamb, W.F., Callaghan, M.W., Fuss, S., Hilaire, J., Creutzig, F., Amann, T., Beringer, T., Garcia, W. de O., Hartmann, J., Khanna, T., Lenzi, D., Luderer, G., Nemet, G.F., Rogelj, J., Smith, P., Vicente, J.L.V., Wilcox, J., Dominguez, M. del M.Z., 2018. Negative emissions—part 1: research landscape and synthesis. Environ. Res. Lett. 13, 063001. https://doi.org/10.1088/1748-9326/aabf9b

Moraes, J.L., Cerri, C.C., Melillo, J.M., Kicklighter, D., Neill, C., Skole, D.L., Steudler, P.A., 1995. Soil carbon stocks of the Brazilian Amazon Basin. Soil Sci. Soc. Am. J. 59, 244–247. https://doi.org/10.2136/sssaj1995.03615995005900010038x

Moreno, J.M., Oechel, W.C., 1992. Factors controlling postfire seedling establishment in southern California chaparral. Oecologia 90, 50–60. https://doi.org/10.1007/BF00317808

Moreno, J.M., Oechel, W.C., 1991. Fire intensity and herbivory effects on postfire resprouting of Adenostoma fasciculatum in southern California chaparral. Oecologia 85, 429–433. https://doi.org/10.1007/BF00320621

Murray, B.C., Sohngen, B., Ross, M.T., 2007. Economic consequences of consideration of permanence, leakage and additionality for soil carbon sequestration projects. Clim. Change 80, 127–143. https://doi.org/10.1007/s10584-006-9169-4

Murty, D., Kirschbaum, M.U.F., Mcmurtrie, R.E., Mcgilvray, H., 2002. Does conversion of forest to agricultural land change soil carbon and nitrogen? a review of the literature. Glob. Change Biol. 8, 105–123. https://doi.org/10.1046/j.1354-1013.2001.00459.x

Myhre, G., Shindell, D., Bréon, F.-M., Collins, W., Fuglestvedt, J., Huang, J., Koch, D., Lamarque, J.-F., Lee, D., Mendoza, B., Nakajima, T., Robock, A., Stephens, G., Zhang, H., 2013. Anthropogenic and natural radiative forcing, in: Climate Change 2013: The Physical Science Basis. Contribution of Working Group I to the Fifth Assessment Report of the Intergovernmental Panel on Climate Change [Stocker, T.F., D. Qin, G.-K. Plattner, M. Tignor, S.K. Allen, J. Boschung, A. Nauels, Y. Xia, V. Bex and P.M. Midgley (Eds.)]. Cambridge University Press, New York, NY, p. 82.

Nolan, M., Dewees, S., Lucero, S., 2021. Identifying effective restoration approaches to maximize plant establishment in California grasslands through a meta‐analysis. Restor. Ecol. 29. https://doi.org/10.1111/rec.13370

Nowak, D.J., 1994. Atmospheric carbon dioxide reduction by Chicago's urban forest. Chic. Urban For. Ecosyst. Results Chic. Urban For. Clim. Proj. 83–94.

[NRCS] National Resource Conservation Service, 2020. Conservation Practice Standard: Soil Carbon Amendment Code 808 (No. 808- CPS-1).

Ohsowski, B.M., Klironomos, J.N., Dunfield, K.E., Hart, M.M., 2012. The potential of soil amendments for restoring severely disturbed grasslands. Appl. Soil Ecol. 60, 77–83.

Opperman, J.J., Merenlender, A.M., 2004. The Effectiveness of Riparian Restoration for Improving Instream Fish Habitat in Four Hardwood-Dominated California Streams. North Am. J. Fish. Manag. 24, 822–834. https://doi.org/10.1577/M03-147.1

Owen, J.J., Parton, W.J., Silver, W.L., 2015. Long-term impacts of manure amendments on carbon and greenhouse gas dynamics of rangelands. Glob. Change Biol. 21, 4533–4547. https://doi.org/10.1111/gcb.13044

Pan, W.L., Port, L.E., Xiao, Y., Bary, A.I., Cogger, C.G., 2017. Soil carbon and nitrogen fraction accumulation with long-term biosolids applications. Soil Sci. Soc. Am. J. 81, 1381–1388.

Parsons, R., Jolly, W.M., Hoffman, C., Ottmar, R., 2016. The role of fuels in extreme fire behavior. Synth. Knowl. Extreme Fire Behav. 55.

Perakis, S.S., Kellogg, C.H., 2007. Imprint of oaks on nitrogen availability and δ15N in California grassland-savanna: a case of enhanced N inputs? Plant Ecol. 191, 209–220. https://doi.org/10.1007/s11258-006-9238-9

Pettit, N.E., Naiman, R.J., 2007. Fire in the riparian zone: characteristics and ecological consequences. Ecosystems 10, 673–687. https://doi.org/10.1007/s10021-007-9048-5

Phyoe, W.W., Wang, F., 2019. A review of carbon sink or source effect on artificial reservoirs. Int. J. Environ. Sci. Technol. 16, 2161–2174. https://doi.org/10.1007/s13762-019-02237-2

Pielke, R.A., Avissar, R., 1990. Influence of landscape structure on local and regional climate. Landsc. Ecol. 4, 133–155.

Potthoff, M., Jackson, L.E., Steenwerth, K.L., Ramirez, I., Stromberg, M.R., Rolston, D.E., 2005. Soil biological and chemical properties in restored perennial grassland in California. Restor. Ecol. 13, 61–73. https://doi.org/10.1111/j.1526-100X.2005.00008.x

Pouyat, R.V., Yesilonis, I.D., Nowak, D.J., 2006. Carbon storage by urban soils in the United States. J. Environ. Qual. 35, 1566. https://doi.org/10.2134/jeq2005.0215

Powlson, D.S., Whitmore, A.P., Goulding, K.W., 2011. Soil carbon sequestration to mitigate climate change: a critical re-examination to identify the true and the false. Eur. J. Soil Sci. 62, 42–55.

Preston, D.L., Johnson, P.T.J., 2012. Importance of native amphibians in the diet and distribution of the aquatic gartersnake (Thamnophis atratus) in the San Francisco Bay Area of California. J. Herpetol. 46, 221–227. https://doi.org/10.1670/10-065

Prichard, S.J., Andreu, A.G., Ottmar, R.D., Eberhardt, E., 2019. Fuel Characteristic Classification System (FCCS) Field Sampling and Fuelbed Development Guide (No. PNW-GTR-972). U.S. Department of Agriculture, Forest Service, Pacific Northwest Research Station, Portland, OR. https://doi.org/10.2737/PNW-GTR-972

Qiu, L., Wei, X., Zhang, X., Cheng, J., 2013. Ecosystem carbon and nitrogen accumulation after grazing exclusion in semiarid grassland. PLoS ONE 8, e55433. https://doi.org/10.1371/journal.pone.0055433

Raciti, S.M., Hutyra, L.R., Finzi, A.C., 2012. Depleted soil carbon and nitrogen pools beneath impervious surfaces. Environ. Pollut. 164, 248–251. https://doi.org/10.1016/j.envpol.2012.01.046

Ravi, S., Breshears, D.D., Huxman, T.E., D'Odorico, P., 2010. Land degradation in drylands: Interactions among hydrologic–aeolian erosion and vegetation dynamics. Geomorphology 116, 236–245. https://doi.org/10.1016/j.geomorph.2009.11.023

Realmonte, G., Drouet, L., Gambhir, A., Glynn, J., Hawkes, A., Köberle, A.C., Tavoni, M., 2019. An inter-model assessment of the role of direct air capture in deep mitigation pathways. Nat. Commun. 10, 3277. https://doi.org/10.1038/s41467-019-10842-5

Reeves, M.C., Ryan, K.C., Rollins, M.G., Thompson, T.G., 2009. Spatial fuel data products of the LANDFIRE Project. Int. J. Wildland Fire 18, 250–267. https://doi.org/10.1071/WF08086

Riccardi, C.L., Ottmar, R.D., Sandberg, D.V., Andreu, A., Elman, E., Kopper, K., Long, J., 2007. The fuelbed: a key element of the Fuel Characteristic Classification System. Can. J. For. Res. 37, 2394–2412.

Rincon Consultants, Inc., 2020. Memorandum Detailing GHG Emissions Inventory, Forecast, and Provisional Targets for Livermore Climate Action Plan Update. Oakland, CA.

Russell, W.H., McBride, J.R., 2003. Landscape scale vegetation-type conversion and fire hazard in the San Francisco bay area open spaces. Landsc. Urban Plan. 64, 201–208. https://doi.org/10.1016/S0169-2046(02)00233-5

Rutherford, K.H., Evett, R.R., Hopkinson, P., 2020. Using phytolith analysis to reconstruct prehistoric fire regimes in central coastal California. Int. J. Wildland Fire 29, 832. https://doi.org/10.1071/WF20013

Ryals, R., Eviner, V.T., Stein, C., Suding, K.N., Silver, W.L., 2016. Grassland compost amendments increase plant production without changing plant communities. Ecosphere 7. https://doi.org/10.1002/ecs2.1270

Ryals, R., Hartman, M.D., Parton, W.J., DeLonge, M.S., Silver, W.L., 2015. Long-term climate change mitigation potential with organic matter management on grasslands. Ecol. Appl. 25, 531–545. https://doi.org/10.1890/13-2126.1

Ryals, R., Kaiser, M., Torn, M.S., Berhe, A.A., Silver, W.L., 2014. Impacts of organic matter amendments on carbon and nitrogen dynamics in grassland soils. Soil Biol. Biochem. 68, 52–61. https://doi.org/10.1016/j.soilbio.2013.09.011

Ryals, R., Silver, W.L., 2013. Effects of organic matter amendments on net primary productivity and greenhouse gas emissions in annual grasslands. Ecol. Appl. 23, 46–59. https://doi.org/10.1890/12-0620.1

Saah, D., Battles, J., Gunn, J., Buchholz, T., Schmidt, D., Roller, G., Romsos, S., 2016. Technical Improvements to the Greenhouse Gas (GHG) Inventory for California Forests and Other Lands. California Air Resources Board, Agreement#14-757.

Safford, H.D., Butz, R.J., Bohlman, Michelle Coppoletta, Estes, B.L., Gross, S.E., Merriam, K.E., Meyer, M.D., Molinari, N.A., Wuenschel, A., 2021. Fire ecology of the North American Mediterranean-climate zone. Fire Ecol. Manag. Past Present Future US For. Ecosyst. 337–392. https://doi.org/10.1007/978-3-030-73267-7_9

Salemi, L.F., Groppo, J.D., Trevisan, R., Marcos de Moraes, J., de Paula Lima, W., Martinelli, L.A., 2012. Riparian vegetation and water yield: A synthesis. J. Hydrol. 454–455, 195–202. https://doi.org/10.1016/j.jhydrol.2012.05.061

Sandel, B., Cornwell, W., Ackerly, D., 2012. Mechanisms of vegetation change in coastal California, with an emphasis on the San Francisco Bay Area., in: Climate Change Impacts on California Vegetation: Physiology, Life History, and Ecosystem Change,William K. Cornwell, Stephanie A. Stuart, Aaron Ramirez, Christopher R. Dolanc, James H. Thorne, David D. Ackerly, eds. California Energy Commission.

Schlesinger, W.H., 1986. Changes in soil carbon storage and associated properties with disturbance and recovery, in: Trabalka, J.R., Reichle, D.E. (Eds.), The Changing Carbon Cycle. Springer New York, New York, NY, pp. 194–220. https://doi.org/10.1007/978-1-4757-1915-4_11

Schwilk, D.W., 2003. Flammability is a niche construction trait: canopy architecture affects fire intensity. Am. Nat. 162, 725–733. https://doi.org/10.1086/379351

Seabloom, E.W., Batzer, E., Chase, J.M., Stanley Harpole, W., Adler, P.B., Bagchi, S., Bakker, J.D., Barrio, I.C., Biederman, L., Boughton, E.H., Bugalho, M.N., Caldeira, M.C., Catford, J.A., Daleo, P., Eisenhauer, N., Eskelinen, A., Haider, S., Hallett, L.M., Svala Jónsdóttir, I., Kimmel, K., Kuhlman, M., MacDougall, A., Molina, C.D., Moore, J.L., Morgan, J.W., Muthukrishnan, R., Ohlert, T., Risch, A.C., Roscher, C., Schütz, M., Sonnier, G., Tognetti, P.M., Virtanen, R., Wilfahrt, P.A., Borer, E.T., 2021. Species loss due to nutrient addition increases with spatial scale in global grasslands. Ecol. Lett. 24, 2100–2112. https://doi.org/10.1111/ele.13838

Seavy, N.E., Gardali, T., Golet, G.H., Griggs, F.T., Howell, C.A., Kelsey, R., Small, S.L., Viers, J.H., Weigand, J.F., 2009. Why climate change makes riparian restoration more important than ever: recommendations for practice and research. Ecol. Restor. 27, 330–338. https://doi.org/10.3368/er.27.3.330

[SFPUC] San Francisco Public Utilities Commission, 2017. SFPUC Alameda Creek Watershed Range-land Management Plan, February 2017 Draft.

[SFPUC] San Francisco Public Utilities Commission, 2015. Physical and biological resources. Chapter 3 in SFPUC Alameda Watershed Habitat Conservation Plan.

[SFPUC] San Francisco Public Utilities Commission, Natural Resources and Lands Management Division, 2016. An Overview of Carbon Storage and Watershed Management.

Shaffer, S., Thompson, E., 2015. A New Comparison of Greenhouse Gas Emissions from California Agricultural and Urban Land Uses. American Farmland Trust, Davis, CA.

Silver, W.L., Ryals, R., Eviner, V., 2010. Soil carbon pools in California's annual grassland ecosystems. Rangel. Ecol. Manag. 63, 128–136. https://doi.org/10.2111/REM-D-09-00106.1

Silver, W.L., Vergara, S.E., Mayer, A., 2018. Carbon sequestration and greenhouse gas mitigation potential of composting and soil amendments on California's rangelands. Calif. Nat. Resour. Agency 62.

Sitch, S., Smith, B., Prentice, I.C., Arneth, A., Bondeau, A., Cramer, W., Kaplan, J.O., Levis, S., Lucht, W., Sykes, M.T., Thonicke, K., Venevsky, S., 2003. Evaluation of ecosystem dynamics, plant geography and terrestrial carbon cycling in the LPJ dynamic global vegetation model. Glob. Change Biol. 9, 161–185. https://doi.org/10.1046/j.1365-2486.2003.00569.x

Six, J., Conant, R.T., Paul, E.A., Paustian, K., 2002. Stabilization mechanisms of soil organic matter: Implications for C-saturation of soils. Plant Soil 241, 155–176. https://doi.org/10.1023/A:1016125726789

Smart, D., Carlisle, E., Spencer, R., 2003. Carbon Flow Through Root and Microbial Respiration in Vineyards and Adjacent Oak Woodland Grassland Communities. Report to the Kearney Foundation in Soil Science.

Smith, P., 2016. Soil carbon sequestration and biochar as negative emission technologies. Glob. Change Biol. 22, 1315–1324. https://doi.org/10.1111/gcb.13178

Spiegal, S., Larios, L., Bartolome, J.W., Suding, K.N., 2014. Restoration management for spatially and temporally complex Californian grassland., in Grassland Biodiversity and Conservation in a Changing World. Nova Science Publishers.

Stahlheber, K.A., 2016. The impacts of isolation, canopy size, and environmental conditions on patterns of understory species richness in an oak savanna. Plant Ecol. 217, 825–841. https://doi.org/10.1007/s11258-016-0605-x

Stahlheber, K.A., D'Antonio, C.M., 2014. Do tree canopies enhance perennial grass restoration in California oak savannas? Restor. Ecol. 22, 574–581. https://doi.org/10.1111/rec.12103

Stahlheber, K.A., D'Antonio, C.M., 2013. Using livestock to manage plant composition: A meta-analysis of grazing in California Mediterranean grasslands. Biol. Conserv. 157, 300–308. https://doi.org/10.1016/j.biocon.2012.09.008

Stanford, B., Grossinger, R., Beagle, J., Askevold, R., Leidy, R., Beller, E., Salomon, M., Striplen, C., Whipple, A., 2013. Alameda Creek Watershed Historical Ecology Study. San Francisco Estuary Institute - Aquatic Science Center, Richmond, CA.

Steenwerth, K.L., Jackson, L.E., Carlisle, E.A., Scow, K.M., 2006. Microbial communities of a native perennial bunchgrass do not respond consistently across a gradient of land-use intensification. Soil Biol. Biochem. 38, 1797–1811. https://doi.org/10.1016/j.soilbio.2005.12.005

Stromberg, M.R., Corbin, J.D., D'Antonio, C.M., 2007. California Grasslands Ecology and Management. University of California Press. https://doi.org/10.1525/california/9780520252202.001.0001

Suding, K.N., Collins, S.L., Gough, L., Clark, C., Cleland, E.E., Gross, K.L., Milchunas, D.G., Pennings, S., 2005. Functional- and abundance-based mechanisms explain diversity loss due to N fertilization. Proc. Natl. Acad. Sci. 102, 4387–4392. https://doi.org/10.1073/pnas.0408648102

Sugihara, N.G., Van Wagtendonk, J.W., Fites-Kaufman, J., 2006. Fire as an ecological process, in: Sugihara, N. (Ed.), Fire in California's Ecosystems. University of California Press, pp. 58–74. https://doi.org/10.1525/california/9780520246058.003.0004

Sulman, B.N., Harden, J., He, Y., Treat, C., Koven, C., Mishra, U., O'Donnell, J.A., Nave, L.E., 2020. Land use and land cover affect the depth distribution of soil carbon: insights from a large database of soil profiles. Front. Environ. Sci. 8.

Suttle, K.B., Thomsen, M.A., 2007. Climate change and grassland restoration in California: lessons from six years of rainfall manipulation in a north coast grassland. Madroño, 54(3), pp.225-233.

Swan, A., Easter, M., Chambers, A., Brown, K., Williams, S.A., Creque, J., Wick, J., Paustian, K., 2015. Carbon and Greenhouse Gas Evaluation for NRCS Conservation Practice Planning. A companion report to www.comet-planner.com.

Sweeney, B.W., Newbold, J.D., 2014. Streamside forest buffer width needed to protect stream water quality, habitat, and organisms: a literature review. JAWRA J. Am. Water Resour. Assoc. 50, 560–584.

Tate, K.W., Atwill, E.R., Bartolome, J.W., Nader, G., 2006. Significant Escherichia coli attenuation by vegetative buffers on annual grasslands. J. Environ. Qual. 35, 795–805.

Tian, G., Chiu, C.-Y., Franzluebbers, A.J., Oladeji, O.O., Granato, T.C., Cox, A.E., 2015. Biosolids amendment dramatically increases sequestration of crop residue-carbon in agricultural soils in western Illinois. Appl. Soil Ecol. 85, 86–93. https://doi.org/10.1016/j.apsoil.2014.09.001

Tisdall, J.M., Oades, J.M., 1982. Organic matter and water-stable aggregates in soils. J. Soil Sci. 33, 141–163. https://doi.org/10.1111/j.1365-2389.1982.tb01755.x

Townsend-Small, A., Czimczik, C.I., 2010. Carbon sequestration and greenhouse gas emissions in urban turf: global warming potential of lawns. Geophys. Res. Lett. 37, n/a-n/a. https://doi.org/10.1029/2009GL041675

Tyler, C.M., Odion, D.C., Callaway R.M., 2007. Dynamics of Woody Species in the California Grassland. In California Grasslands: Ecology and Management, eds. Mark Stromberg, Jeffrey Corbin, Carla D'Antonio. Berkeley, CA: University of California Press.

Urbanski, S., 2014. Wildland fire emissions, carbon, and climate: Emission factors. For. Ecol. Manag., Wildland fire emissions, carbon, and climate: Science overview and knowledge needs 317, 51–60. https://doi.org/10.1016/j.foreco.2013.05.045

USDA NRCS, 2016. Conservation Practice Standard Silvopasture Code 381.

U.S. Department of Agriculture (USDA), Western Division Laboratory. 1939-40. [Aerial photos of Alameda County]. Scale: 1:20,000. Agricultural Adjustment Administration (AAA). Courtesy of Earth Sciences & Map Library, UC Berkeley, and the Alameda County Resource Conservation District (ACRCD) and National Resources Conservation Service (NRCS).

van der Werf, G.R., Randerson, J.T., Giglio, L., van Leeuwen, T.T., Chen, Y., Rogers, B.M., Mu, M., van Marle, M.J.E., Morton, D.C., Collatz, G.J., Yokelson, R.J., Kasibhatla, P.S., 2017. Global fire emissions estimates during 1997–2016. Earth Syst. Sci. Data 9, 697–720. https://doi.org/10.5194/essd-9-697-2017

Venkat, K., 2012. Comparison of Twelve Organic and Conventional Farming Systems: A Life Cycle Greenhouse Gas Emissions Perspective. J. Sustain. Agric. 36, 620–649. https://doi.org/10.1080/10440046.2012.672378

Verhoeven, E., Pereira, E., Decock, C., Garland, G., Kennedy, T., Suddick, E., Horwath, W., Six, J., 2017. N2O emissions from California farmlands: A review. Calif. Agric. 71, 148–159.

Villa, Y.B., Ryals, R., 2021. Soil carbon response to long-term biosolids application. J. Environ. Qual. 50, 1084–1096. https://doi.org/10.1002/jeq2.20270

Waldrop, M.P., Firestone, M.K., 2006. Response of microbial community composition and function to soil climate change. Microb. Ecol. 52, 716–724.

Wang, I.J., Savage, W.K., Bradley Shaffer, H., 2009. Landscape genetics and least-cost path analysis reveal unexpected dispersal routes in the California tiger salamander (Ambystoma californiense). Mol. Ecol. 18, 1365–1374.

Weiss, S.B., 1999. Cars, cows, and checkerspot butterflies: nitrogen deposition and management of nutrient-poor grasslands for a threatened species. Conserv. Biol. 13, 1476–1486. https://doi.org/10.1046/j.1523-1739.1999.98468.x

Welsh, M.K., Vidon, P.G., McMillan, S.K., 2021. Riparian seasonal water quality and greenhouse gas dynamics following stream restoration. Biogeochemistry 156, 453–474. https://doi.org/10.1007/s10533-021-00866-9

Whittier, J., Rue, D., Haase, S., 1994. Urban tree residues: results of the first national inventory. (No. CONF-9410176-). Western Regional Biomass Energy Program, Reno, NV (United States).

Wijesekara, H., Bolan, N.S., Thangavel, R., Seshadri, B., Surapaneni, A., Saint, C., Hetherington, C., Matthews, P., Vithanage, M., 2017. The impact of biosolids application on organic carbon and carbon dioxide fluxes in soil. Chemosphere 189, 565–573.

Williams, J.N., Hollander, A.D., O'Geen, A.T., Thrupp, L.A., Hanifin, R., Steenwerth, K., McGourty, G., Jackson, L.E., 2011. Assessment of carbon in woody plants and soil across a vineyard-woodland landscape. Carbon Balance Manag. 6, 11.

Williams, K., Hobbs, R.J., Hamburg, S.P., 1987. Invasion of an annual grassland in Northern California by Baccharis pilularis ssp. consanguinea. Oecologia 72, 461–465. https://doi.org/10.1007/BF00377580

Wilsey, B.J., Wayne Polley, H., 2006. Aboveground productivity and root–shoot allocation differ between native and introduced grass species. Oecologia 150, 300–309.

Xu, L., Baldocchi, D.D., 2004. Seasonal variation in carbon dioxide exchange over a Mediterranean annual grassland in California. Agric. For. Meteorol. 123, 79–96. https://doi.org/10.1016/j.agrformet.2003.10.004

Yan, Y., Kuang, W., Zhang, C., Chen, C., 2015. Impacts of impervious surface expansion on soil organic carbon – a spatially explicit study. Sci. Rep. 5, 1–9. https://doi.org/10.1038/srep17905

Zavaleta, E.S., Kettley, L.S., 2006. Ecosystem change along a woody invasion chronose-quence in a California grassland. J. Arid Environ. 66, 290–306. https://doi.org/10.1016/j.jaridenv.2005.11.008

Zefferman, E., 2014. Increasing canopy shading reduces growth but not establishment of Elodea nuttallii and Myriophyllum spicatum in stream channels. Hydrobiologia 734, 159–170.

www.ingramcontent.com/pod-product-compliance
Lightning Source LLC
Chambersburg PA
CBHW061149030426
42335CB00003B/156